Recent Developments of Nanofluids

Special Issue Editor
Rahmat Ellahi

MDPI • Basel • Beijing • Wuhan • Barcelona • Belgrade

Special Issue Editor
Rahmat Ellahi
University Islamabad (IIUI)
Pakistan

Editorial Office
MDPI
St. Alban-Anlage 66
Basel, Switzerland

This edition is a reprint of the Special Issue published online in the open access journal *Applied Sciences* (ISSN 2076-3417) from 2016–2018 (available at: http://www.mdpi.com/journal/applsci/special_issues/nanofluids).

For citation purposes, cite each article independently as indicated on the article page online and as indicated below:

Lastname, F.M.; Lastname, F.M. Article title. *Journal Name* **Year**, *Article number*, page range.

First Edition 2018

ISBN 978-3-03842-833-6 (Pbk)
ISBN 978-3-03842-834-3 (PDF)

Table of Contents

About the Special Issue Editor

Rahmat Ellahi, Ph.D., Professor, Ex-Chairperson (Mathematics and Statistics), accomplished researcher, teacher and a prolific scholar has key role in promotion of science at national and international levels. He has successfully achieved great height of academics, from the University of Punjab, Quaid-i-Azam University, Islamabad, Pakistan and University of California Riverside, USA as his Alma mater. He has published around 200 papers in the journals of USA, Germany, UK, Canada etc. His research particularly upgraded science capacity of several Universities of the world such as USA, Canada, Romania, Iran, South Africa, Saudi Arabia and Pakistan etc. His work has been cited more than 7700 times at Google Scholar having 48 h-index. He is an author of 06 books published at national and international levels. Besides, he edited 03 special issues for ISI impact factor journals.

He is editor/editorial board member for 14 international journals and referee for more than 235 international journals. Approved Supervisor of Higher Education Commission of Pakistan (HEC) for MS/M. Phil/Ph.D. Students since 2006. He has successfully supervised 23 research students (06 Ph.D. and 17 MS). His leadership in academics is further reflected through the research collaboration with more than 50 international leading scientists all over the world such as USA, Canada, South Africa, India, Saudi Arabia Iran, Turkey and Romania etc.

He has been continuously honored with annual Research Productivity Award based on his excellent scientific achievements by PCST, Ministry of Science and Technology since 2009 to date. PCST has honed him with 5th top most Productive Scientist of Pakistan Award "Category A" in consecutive three years.

He has organized 08 international conferences; delivered 20 seminars and attended 25 conferences as key speaker and participant. As a referee he has investigated 33 research projects submitted at HEC and USEFP under NRUP and Fulbright Grant (for Pakistan and Poland Scholars).

He has received 05 (03 international and 04 national) awards and several honors. He is actively involved in different professional and academia bodies/institutes at national and international levels. He also established strong research collaboration with more than 50 prominent researchers. Dr. Ellahi has established 04 new departments; Mathematic, Statistics, Physics and Quality Enhancement Cell at IIUI Pakistan. In summary, Dr. Rahmat Ellahi appears to be a superb individual encompassing all facets of a great educationist. As a matter of fact he is an outstanding candidate and it is because of his remarkable contributions towards society, teaching, research, development and promotion of scientific cooperation, international collaboration, human resource development and updating education systems with the latest trends.

Preface to "Recent Developments of Nanofluids"

This book contains nine chapters. First eight chapters have independent strength and point of emphasis depending on the pen of authors whereas Editorial is given in chapter nine. Extensive uses of realistic applications are commonly given in each chapter. For the best understanding of readers, a relevant list of references is also given at the end of each chapter for further study.

I wish to thank excellent reviewers for their suggestions and critical reviews on submitted manuscripts. I was fortunate enough to have prominent scholars who contributed with their original research work. I applaud all of them on successful completion of this book.

Errors and omissions if any are requested to point out which will be gratefully acknowledged in the next possible Edition. Particularly, suggestions for improvement, scope and format of the book will be highly appreciated.

I express my gratitude to MPDI for publishing this book especially I also want to express my gratitudes to Ms. Jennifer Li, my family and friends for their helpful cooperation.

Rahmat Ellahi

Special Issue Editor

Article

On Squeezed Flow of Jeffrey Nanofluid between Two Parallel Disks

Tasawar Hayat [1,2], Tehseen Abbas [1,*], Muhammad Ayub [1], Taseer Muhammad [1] and Ahmed Alsaedi [2]

1 Department of Mathematics, Quaid-I-Azam University, Islamabad 44000, Pakistan; fmgpak@gmail.com (T.H.); mayub@qau.edu.pk (M.A.); taseer@math.qau.edu.pk (T.M.)
2 Nonlinear Analysis and Applied Mathematics (NAAM) Research Group, Department of Mathematics, Faculty of Science, King Abdulaziz University, Jeddah 21589, Saudi Arabia; aalsaedi@hotmail.com
* Correspondence: tehseen@math.qau.edu.pk; Tel.: +92-51-9064-2172

Academic Editor: Rahmat Ellahi
Received: 7 September 2016 ; Accepted: 31 October 2016 ; Published: 11 November 2016

Abstract: The present communication examines the magnetohydrodynamic (MHD) squeezing flow of Jeffrey nanofluid between two parallel disks. Constitutive relations of Jeffrey fluid are employed in the problem development. Heat and mass transfer aspects are examined in the presence of thermophoresis and Brownian motion. Jeffrey fluid subject to time dependent applied magnetic field is conducted. Suitable variables lead to a strong nonlinear system. The resulting systems are computed via homotopic approach. The behaviors of several pertinent parameters are analyzed through graphs and numerical data. Skin friction coefficient and heat and mass transfer rates are numerically examined.

Keywords: squeezing flow; Jeffrey fluid; nanoparticles; magnetic field

1. Introduction

The homogenous mixture of ultrafine nanometer-sized particles and convectional heat transfer base liquids is termed as nanofluid. Nanomaterials have a key role in the industrial and engineering processes like processing of coolants for the nuclear reactors, transformer coolant and radiation therapy in cancer treatment etc. Furthermore, the magneto-nanofluid is very helpful in various sectors including sterilizing devices, oil recovery from the underground reservoirs, gastric medications, and tumor elimination with hyperthermia. The small sized nanoparticles (which are mostly metallic, nonmetallic, metal-oxides) are good thermal conductors. For this reason, the nanofluid in comparison to the base fluid has greater thermal efficiency. Choi [1] proposed the idea of nanofluid. He argued that the addition of nanoparticles into the base fluid enhances the thermal performance of base fluid. Buongiorno [2] provided expressions including thermophoresis and Brownian motion. Later on, numerous researchers discussed the flows of nanofluid under different geometries. The relevant literature can be seen through the investigations [3–20] and several studies therein.

Squeezing flow between the parallel disks has received the attention of recent researchers due to widespread applications of such flows in various mechanical engineering disciplines. The flow is generated because of two parallel approaching surfaces in relative motion. The parallel approaching surfaces phenomena along with the relative motion is mostly used by the engineers in the modeling of flow of oil in bearings, determination of capacity of load-bearings, compression and injection modeling, etc. (see [21,22]). Stefan [23] reported the squeezing flow for lubrication approximation. Domairry and Aziz [24] studied magnetohydrodynamic squeezing flow of viscous liquid bounded by parallel disks. Siddiqui et al. [25] examined squeezing flow subject to an applied magnetic field. Rashidi et al. [26]

performed an analysis of hydrodynamic squeezing flow by developing series solutions. Some other investigations on squeezing flow can be seen in the studies [27–30].

The prime interest in the present communication is to venture further into the regime of the squeezing flow of non-Newtonian nanofluid. Therefore, the explicit contribution here is as follows: firstly, to formulate the relevant problem for constitutive relations of the Jeffrey fluid model; secondly, to analyze Brownian motion and thermophoresis; thirdly, to consider magnetohydrodynamics of nanofluid; and fourthly, to entertain the idea of permeable characteristics of lower disks. The upper impermeable disk moves towards the lower disk with time-dependent velocity. Problem formulation is made through small magnetic Reynolds number approximation. The homotopy analysis technique (HAM) [31–40] is applied to obtain the convergent solutions of the governing equations. The present study has been arranged as follows. The next section presents problem development. Section 3 depicts the development convergent series solutions. Analysis for convergence and discussion have been examined in Sections 4 and 5, respectively. Section 6 gives the main outcomes of the present study. Note that the considered Jeffrey fluid, although capturing the salient features of relaxation and retardation time, is not able to predict the shear thinning/shear thickening and normal stress effects.

2. Formulation

Consider magnetohydrodynamic squeezing flow of a Jeffrey nanofluid between the two parallel disks. The distance between the parallel disks is $h(t) = H(1-\alpha t)^{1/2}$. The upper disk is at $z = h(t)$, whereas the lower permeable disk is at $z = 0$. A magnetic field $B(t) = B_0(1-\alpha t)^{-1/2}$ is taken transverse to the flow. Here, the induced magnetic field is neglected for a small magnetic Reynolds number [41–43]. Brownian motion and thermophoresis phenomena are accounted. The governing equations for Jeffrey nanofluid are differences in traffic flow

$$\frac{\partial u}{\partial r} + \frac{u}{r} + \frac{\partial w}{\partial z} = 0, \tag{1}$$

$$\begin{aligned}\left(\frac{\partial u}{\partial t} + u\frac{\partial u}{\partial r} + w\frac{\partial u}{\partial z}\right) &= -\frac{1}{\rho}\frac{\partial p}{\partial r} + \frac{\nu}{1+\lambda_1}\left(\frac{\partial^2 u}{\partial r^2} + \frac{\partial^2 u}{\partial z^2} + \frac{1}{r}\frac{\partial u}{\partial r} - \frac{u}{r^2}\right) \\ &+ \frac{\nu\lambda_2}{1+\lambda_1}\begin{pmatrix} \frac{\partial^3 u}{\partial t\partial z^2} + 2\frac{\partial^3 u}{\partial t\partial r^2} + \frac{2}{r}\frac{\partial^2 u}{\partial t\partial z} + \frac{\partial^3 w}{\partial t\partial r\partial z} \\ -\frac{2}{r^2}\frac{\partial u}{\partial t} + \frac{\partial u}{\partial r}\left(\frac{\partial^2 u}{\partial r^2} - 2\frac{u}{r^2}\right) + \frac{\partial w}{\partial r}\frac{\partial^2 u}{\partial z\partial r} \\ +u\left(\frac{\partial^3 u}{\partial r^3} + \frac{\partial^3 w}{\partial r^2\partial z} + \frac{\partial^3 u}{\partial r\partial z^2} + 2\frac{u}{r^3}\right) \\ +w\left(\frac{\partial^3 u}{\partial z\partial r^2} + \frac{2}{r}\frac{\partial^2 u}{\partial z\partial r} + \frac{\partial^3 w}{\partial z^3} + \frac{\partial^2 w}{\partial r\partial z}\right) \\ +\frac{\partial u}{\partial z}\left(\frac{\partial^2 w}{\partial r^2} - 2\frac{w}{r^2} + \frac{\partial^2 u}{\partial z\partial r}\right) \\ +\frac{\partial w}{\partial z}\left(\frac{\partial^2 u}{\partial z^2} + \frac{\partial^2 w}{\partial r\partial z}\right) + \frac{2}{r}\frac{\partial^2 u}{\partial r^2} \end{pmatrix} - \frac{\sigma B^2}{\rho}u,\end{aligned} \tag{2}$$

$$\begin{aligned}\left(\frac{\partial w}{\partial t} + u\frac{\partial w}{\partial r} + w\frac{\partial w}{\partial z}\right) &= -\frac{1}{\rho}\frac{\partial p}{\partial z} + \frac{\nu}{1+\lambda_1}\left(\frac{\partial^2 w}{\partial r^2} + \frac{\partial^2 w}{\partial z^2} + \frac{1}{r}\frac{\partial w}{\partial r}\right) \\ &+ \frac{\nu\lambda_2}{1+\lambda_1}\begin{pmatrix} \frac{\partial^3 w}{\partial r\partial t\partial z} + \frac{\partial^3 w}{\partial t\partial r^2} + 2\frac{\partial^3 u}{\partial r\partial z^2} \\ +2\frac{\partial u}{\partial z}\frac{\partial^2 w}{\partial r\partial z} + \frac{\partial u}{\partial r}\left(\frac{\partial^2 u}{\partial r\partial z} + \frac{\partial^2 w}{\partial r^2}\right) \\ +u\left(\frac{\partial^3 w}{\partial r^2\partial z} + \frac{\partial^3 u}{\partial r^3}\right) + \frac{\partial w}{\partial r}\left(\frac{\partial^2 u}{\partial z^2} + \frac{\partial^3 u}{\partial t\partial z^2}\right) \\ +w\left(\frac{\partial^3 u}{\partial r\partial z^2} + \frac{\partial^3 w}{\partial r^2\partial z} + \frac{\partial^3 w}{\partial z^3} + \frac{1}{r}\frac{\partial^2 u}{\partial z^2} + \frac{1}{r}\frac{\partial^2 w}{\partial z\partial r}\right) \\ +\frac{u}{r}\left(\frac{\partial^2 u}{\partial r\partial z} + \frac{\partial^2 w}{\partial r^2}\right) + \frac{1}{r}\left(\frac{\partial^2 u}{\partial t\partial z} + \frac{\partial^2 w}{\partial t\partial r}\right) \end{pmatrix},\end{aligned} \tag{3}$$

$$\frac{\partial T}{\partial t} + u\frac{\partial T}{\partial r} + w\frac{\partial T}{\partial z} = \alpha\left(\frac{\partial^2 T}{\partial r^2} + \frac{1}{r}\frac{\partial T}{\partial r} + \frac{\partial^2 T}{\partial z^2}\right) + \tau\begin{pmatrix} D_B\left(\frac{\partial C}{\partial r}\frac{\partial T}{\partial r} + \frac{\partial C}{\partial z}\frac{\partial T}{\partial z}\right) \\ +\frac{D_T}{T_m}\left(\left(\frac{\partial T}{\partial r}\right)^2 + \left(\frac{\partial T}{\partial z}\right)^2\right) \end{pmatrix}, \tag{4}$$

$$\frac{\partial C}{\partial t} + u\frac{\partial C}{\partial r} + w\frac{\partial C}{\partial z} = D_B\left(\frac{\partial^2 C}{\partial r^2} + \frac{1}{r}\frac{\partial C}{\partial r} + \frac{\partial^2 C}{\partial z^2}\right) + \frac{D_T}{T_m}\left(\frac{\partial^2 T}{\partial r^2} + \frac{1}{r}\frac{\partial T}{\partial r} + \frac{\partial^2 T}{\partial z^2}\right), \tag{5}$$

with the associated boundary conditions

$$\left.\begin{array}{l} u=0,\ w=-w_0,\ T=T_w,\ C=C_w \text{ at } z=0, \\ u=0,\ w=\frac{\partial h}{\partial t},\ T=T_h,\ C=C_h \text{ at } z=h(t) \end{array}\right\}. \tag{6}$$

Here, u and w denote the velocity components along the r- and z- directions, respectively, p the pressure, $\nu=\mu/\rho$ the kinematic viscosity, μ the dynamic viscosity, ρ the density of base fluid, σ the electrical conductivity, λ_1 the ratio of relaxation and retardation times, λ_2 the retardation time, respectively, T the temperature, $\tau=(\rho c)_p/(\rho c)_f$ the ratio of effective heat capacity of nanoparticles and heat capacity of fluid, $(\rho c)_p$ the effective heat capacity of nanoparticles, $(\rho c)_f$ the heat capacity of fluid, C the concentration, T_m the mean fluid temperature, $\alpha=k/(\rho c)_f$ the thermal diffusivity, k the thermal conductivity, D_B the Brownian diffusion coefficient and D_T the thermophoresis diffusion coefficient. Consider

$$u=\frac{\alpha r}{2(1-\alpha t)}f'(\eta),\ w=-\frac{\alpha H}{\sqrt{1-\alpha t}}f(\eta),\ \eta=\frac{z}{H\sqrt{1-\alpha t}}, \tag{7}$$

$$\theta(\eta)=\frac{T-T_h}{T_w-T_h},\ \phi(\eta)=\frac{C-C_h}{C_w-C_h}. \tag{8}$$

Equations (2)–(6) after elimination of pressure gradient yield

$$f^{iv}-Sq(1+\lambda_1)(\eta f'''+3f''-2ff''')+\frac{\beta}{2}(\eta f^{v}+5f^{iv}+f''f'''-3f'f^{iv})-M^2(1+\lambda_1)f''=0, \tag{9}$$

$$\theta''+PrSq(f\theta'-\eta\theta')+PrNb\theta'\phi'+PrNt\theta'^2=0, \tag{10}$$

$$\phi''+PrLeSq(f\phi'-\eta\phi')+\frac{Nt}{Nb}\theta''=0, \tag{11}$$

$$f(0)=S,\ f'(0)=0,\ \theta(0)=1,\ \phi(0)=1, \tag{12}$$

$$f(1)=\frac{1}{2},\ f'(1)=0,\ \theta(1)=0,\ \phi(1)=0. \tag{13}$$

Here, Pr denotes the Prandtl number, Le the Lewis number, Nb Brownian motion parameter, S the suction/blowing parameter, Nt the thermophoresis parameter, β the Deborah number, M the Hartman number and Sq the squeezing parameter. These quantities are expressed as follows:

$$\begin{array}{lll} Pr &=& \frac{\nu}{\alpha},\ Le=\frac{\alpha}{D_B},\ Nb=\frac{\tau D_B}{\nu}(C_w-C_h),\ S=\frac{w_0}{\alpha H}, \\ Sq &=& \frac{\alpha H^2}{2\nu},\ Nt=\frac{\tau D_T}{\nu T_m}(T_w-T_h),\ M=HB_0\sqrt{\frac{\sigma}{\mu}},\ \beta=\frac{\lambda_2\alpha}{1-\alpha t}. \end{array} \tag{14}$$

Expressions of skin frictions corresponding to lower and upper disks are

$$C_{f1}=\frac{\tau_{rz}|_{z=0}}{\rho\left(\frac{\alpha H}{2(1-\alpha t)^{1/2}}\right)^2}, \tag{15}$$

and

$$C_{f2}=\frac{\tau_{rz}|_{z=h(t)}}{\rho\left(\frac{\alpha H}{2(1-\alpha t)^{1/2}}\right)^2}, \tag{16}$$

with

$$\begin{array}{lll} \tau_{rz} &=& \frac{\mu}{1+\lambda_1}\left(\frac{\partial u}{\partial z}+\frac{\partial w}{\partial r}\right) \\ && +\frac{\lambda_2}{1+\lambda_1}\left(\frac{\partial^2 u}{\partial t\partial z}+\frac{\partial^2 w}{\partial t\partial r}+u\left(\frac{\partial^2 u}{\partial r\partial z}+\frac{\partial^2 w}{\partial r^2}\right)+w\left(\frac{\partial^2 u}{\partial z^2}+\frac{\partial^2 w}{\partial z\partial r}\right)\right). \end{array} \tag{17}$$

The dimensionless forms of skin friction coefficients are

$$\frac{H^2}{r^2}Re_r C_{f1} = \left(1+\frac{3}{2}\beta\right) f''(0), \tag{18}$$

and

$$\frac{H^2}{r^2}Re_r C_{f2} = \left(1+\frac{3}{2}\beta\right) f''(1), \tag{19}$$

where

$$Re_r^{-1} = \frac{2\nu}{r\alpha H(1+\lambda_1)(1-\alpha t)^{1/2}}. \tag{20}$$

Local Nusselt numbers at lower and upper disks are given by

$$Nu_{r1} = -\frac{H}{(T_w - T_h)}\left.\frac{\partial T}{\partial z}\right|_{z=0} = -\frac{1}{\sqrt{1-\alpha t}}\theta'(0), \tag{21}$$

and

$$Nu_{r2} = -\frac{H}{(T_w - T_h)}\left.\frac{\partial T}{\partial z}\right|_{z=h(t)} = -\frac{1}{\sqrt{1-\alpha t}}\theta'(1). \tag{22}$$

Local Sherwood numbers at lower and upper disks can be expressed as follows:

$$Sh_{r_1} = -\frac{H}{(C_w - C_h)}\left.\frac{\partial C}{\partial z}\right|_{z=0} = -\frac{1}{\sqrt{1-\alpha t}}\phi'(0), \tag{23}$$

and

$$Sh_{r_2} = -\frac{H}{(C_w - C_h)}\left.\frac{\partial C}{\partial z}\right|_{z=h(t)} = -\frac{1}{\sqrt{1-\alpha t}}\phi'(1). \tag{24}$$

3. Homotopic Solutions

3.1. Zeroth-Order Deformation

Here, we construct the convergent series solutions of the incoming nonlinear systems. For these, the initial approximation and auxiliary linear operators are taken in the form

$$f_0(\eta) = (-1+2S)\eta^3 - \frac{1}{2}(-3+6S)\eta^2 + S,\ \theta_0(\eta) = 1-\eta,\ \phi_0(\eta) = 1-\eta, \tag{25}$$

$$\mathcal{L}_f(f) = \frac{d^4 f}{d\eta^4},\ \mathcal{L}_\theta(\theta) = \frac{d^2\theta}{d\eta^2},\ \mathcal{L}_\phi(\phi) = \frac{d^2\phi}{d\eta^2}, \tag{26}$$

with the properties

$$\mathcal{L}_f\left[B_1^* + B_2^*\eta + B_3^*\eta^2 + B_4^*\eta^3\right] = 0, \tag{27}$$

$$\mathcal{L}_\theta\left[B_5^* + B_6^*\eta\right] = 0, \tag{28}$$

$$\mathcal{L}_\phi\left[B_7^* + B_8^*\eta\right] = 0. \tag{29}$$

Here, B_i^* $(i = 1-8)$ are the arbitrary constants. The zeroth-order deformation statements are

$$(1-\text{Þ})\mathcal{L}_f\left[\widehat{f}(\eta;\text{Þ}) - f_0(\eta)\right] = \text{Þ}\hbar_f\mathcal{N}_f\left[\widehat{f}(\eta;\text{Þ})\right], \tag{30}$$

$$(1-\text{Þ})\mathcal{L}_\theta\left[\widehat{\theta}(\eta;\text{Þ}) - \theta_0(\eta)\right] = \text{Þ}\hbar_\theta\mathcal{N}_\theta\left[\widehat{f}(\eta;\text{Þ}), \widehat{\theta}(\eta;\text{Þ}), \widehat{\phi}(\eta;\text{Þ})\right], \tag{31}$$

$$(1-\text{Þ})\mathcal{L}_\phi\left[\widehat{\phi}(\eta;\text{Þ}) - \phi_0(\eta)\right] = \text{Þ}\hbar_\phi\mathcal{N}_\phi\left[\widehat{f}(\eta;\text{Þ}), \widehat{\theta}(\eta;\text{Þ}), \widehat{\phi}(\eta;\text{Þ})\right], \tag{32}$$

$$\widehat{f}(0;\text{Þ}) = S,\ \widehat{f}'(0;\text{Þ}) = 0,\ \widehat{\theta}(0;\text{Þ}) = 1,\ \widehat{\phi}(0;\text{Þ}) = 1, \tag{33}$$

$$\widehat{f}(1;\text{Þ}) = \frac{1}{2},\ \widehat{f}'(1;\text{Þ}) = 0,\ \widehat{\theta}(1;\text{Þ}) = 0,\ \widehat{\phi}(1;\text{Þ}) = 0, \tag{34}$$

$$\begin{aligned}\mathcal{N}_f\left[\widehat{f}(\eta,\text{Þ})\right] &= \frac{\partial^4\widehat{f}(\eta;\text{Þ})}{\partial\eta^4} - Sq(1+\lambda_1)\left(\begin{array}{c}\eta\frac{\partial^3\widehat{f}(\eta;\text{Þ})}{\partial\eta^3} + 3\frac{\partial^2\widehat{f}(\eta;\text{Þ})}{\partial\eta^2}\\ -2\frac{\partial\widehat{f}(\eta;\text{Þ})}{\partial\eta}\frac{\partial^3\widehat{f}(\eta;\text{Þ})}{\partial\eta^3}\end{array}\right)\\ &+\frac{\beta}{2}\left(\begin{array}{c}\eta\frac{\partial^5\widehat{f}(\eta;\text{Þ})}{\partial\eta^5} + 5\frac{\partial^4\widehat{f}(\eta;\text{Þ})}{\partial\eta^4}\\ +\frac{\partial^2\widehat{f}(\eta;\text{Þ})}{\partial\eta^2}\frac{\partial^3\widehat{f}(\eta;\text{Þ})}{\partial\eta^3} - 3\frac{\partial\widehat{f}(\eta;\text{Þ})}{\partial\eta}\frac{\partial^4\widehat{f}(\eta;\text{Þ})}{\partial\eta^4}\end{array}\right) - M^2(1+\lambda_1)\frac{\partial^2\widehat{f}(\eta;\text{Þ})}{\partial\eta^2},\end{aligned} \tag{35}$$

$$\begin{aligned}\mathcal{N}_\theta\left[\widehat{f}(\eta,\text{Þ}),\widehat{\theta}(\eta;\text{Þ}),\widehat{\phi}(\eta;\text{Þ})\right] &= \frac{1}{Pr}\frac{\partial^2\widehat{\theta}(\eta;\text{Þ})}{\partial\eta^2} + Sq\left(\widehat{f}(\eta;\text{Þ})\frac{\partial\widehat{\theta}(\eta;\text{Þ})}{\partial\eta} - \eta\frac{\partial\widehat{\theta}(\eta;\text{Þ})}{\partial\eta}\right)\\ &+Nb\frac{\partial\widehat{\theta}(\eta;\text{Þ})}{\partial\eta}\frac{\partial\widehat{\phi}(\eta;\text{Þ})}{\partial\eta} + Nt\left(\frac{\partial\widehat{\theta}(\eta;\text{Þ})}{\partial\eta}\right)^2,\end{aligned} \tag{36}$$

$$\begin{aligned}\mathcal{N}_\phi\left[\widehat{f}(\eta,\text{Þ}),\widehat{\theta}(\eta;\text{Þ}),\widehat{\phi}(\eta;\text{Þ})\right] &= \frac{\partial^2\widehat{\phi}(\eta;\text{Þ})}{\partial\eta^2} + PrLeSq\widehat{f}(\eta;\text{Þ})\frac{\partial\widehat{\phi}(\eta;\text{Þ})}{\partial\eta}\\ &-PrLeSq\eta\frac{\partial\widehat{\phi}(\eta;\text{Þ})}{\partial\eta} + \frac{Nt}{Nb}\frac{\partial^2\widehat{\theta}(\eta;\text{Þ})}{\partial\eta^2},\end{aligned} \tag{37}$$

Here, Þ$\in$ $[0,1]$ indicates the embedding parameter and $\hbar_f$, $\hbar_\theta$ and $\hbar_\phi$ the non-zero auxiliary parameters.

3.2. m^{th}-Order Deformation Equations

$$\mathcal{L}_f\left[f_m(\eta) - \chi_m f_{m-1}(\eta)\right] = \hbar_f\mathcal{R}_m^f(\eta), \tag{38}$$

$$\mathcal{L}_\theta\left[\theta_m(\eta) - \chi_m\theta_{m-1}(\eta)\right] = \hbar_\theta\mathcal{R}_m^\theta(\eta), \tag{39}$$

$$\mathcal{L}_\phi\left[\phi_m(\eta) - \chi_m\phi_{m-1}(\eta)\right] = \hbar_\phi\mathcal{R}_m^\phi(\eta), \tag{40}$$

$$f_m(0) = f_m'(0) = 0,\ \theta_m(0) = \phi_m(0) = 0, \tag{41}$$

$$f_m(1) = f_m'(1) = 0,\ \phi_m(1) = \theta_m(1) = 0, \tag{42}$$

$$\begin{aligned}\mathcal{R}_m^f(\eta) &= f_{m-1}^{iv} - Sq(1+\lambda_1)\left(\eta f_{m-1}''' + 3f_{m-1}'' - 2\textstyle\sum_{k=0}^{m-1} f_{m-1-k}f_k'''\right)\\ &+\frac{\beta}{2}\left(\eta f_{m-1}^{v} + 5f_{m-1}^{iv} + \textstyle\sum_{k=0}^{m-1} f_{m-1-k}''f_k''' - 3\sum_{k=0}^{m-1} f_{m-1-k}'f_k^{iv}\right) - M^2(1+\lambda_1)f_{m-1}'',\end{aligned} \tag{43}$$

$$\mathcal{R}_m^\theta(\eta) = \frac{1}{Pr}\theta_{m-1}'' + Sq\left(\sum_{k=0}^{m-1} f_{m-1-k}\theta_k' - \eta\theta_{m-1}'\right) + Nb\sum_{k=0}^{m-1}\theta_{m-1-k}'\phi_k' + Nt\sum_{k=0}^{m-1}\theta_{m-1-k}'\theta_k', \tag{44}$$

$$\mathcal{R}_m^\phi(\eta) = \phi_{m-1}'' + \frac{Nt}{Nb}\theta_{m-1}'' + PrLeSq\left(\sum_{k=0}^{m-1} f_{m-1-k}\phi_k' - \eta\phi_{m-1}'\right), \tag{45}$$

$$\chi_m = \begin{cases}0, & m\le 1,\\ 1, & m>1.\end{cases} \tag{46}$$

The general solutions (f_m, θ_m, ϕ_m) consisting of the special solutions $(f_m^*, \theta_m^*, \phi_m^*)$ are

$$f_m(\eta) = f_m^*(\eta) + B_1^* + B_2^*\eta + B_3^*\eta^2 + B_4\eta^3, \tag{47}$$

$$\theta_m(\eta) = \theta_m^*(\eta) + B_5^* + B_6^*\eta, \tag{48}$$

$$\phi_m(\eta) = \phi_m^*(\eta) + B_7^* + B_8^*\eta, \tag{49}$$

where the constants B_i^* $(i = 1 - 8)$ are computed through the boundary conditions (41) and (42) with values

$$\left.\begin{aligned} B_1^* &= f_m^*(\eta)\big|_{\eta=0},\ B_2^* = \frac{\partial f_m^*(\eta)}{\partial \eta}\bigg|_{\eta=0}, \\ B_3^* &= -3\, f_m^*(\eta)\big|_{\eta=1} + \frac{\partial f_m^*(\eta)}{\partial \eta}\bigg|_{\eta=1} - 3B_1^* - 2B_2^*, \\ B_4^* &= 2\, f_m^*(\eta)\big|_{\eta=1} - \frac{\partial f_m^*(\eta)}{\partial \eta}\bigg|_{\eta=1} + 2B_1^* + B_2^*, \\ B_5^* &= -\,\theta_m^*(\eta)\big|_{\eta=0},\ B_6^* = \theta_m^*(\eta)\big|_{\eta=0} - \theta_m^*(\eta)\big|_{\eta=1}, \\ B_7^* &= -\,\phi_m^*(\eta)\big|_{\eta=0},\ B_8^* = \phi_m^*(\eta)\big|_{\eta=0} - \phi_m^*(\eta)\big|_{\eta=1}. \end{aligned}\right\} \tag{50}$$

4. Convergence Analysis

Clearly, the approximate series solutions involve the nonzero auxiliary parameters $\hbar_f$, $\hbar_\theta$ and $\hbar_\phi$. To get the appropriate values of $\hbar_f$, $\hbar_\theta$ and $\hbar_\phi$, the $\hbar-$curves are plotted at 20th order of deformations. Figures 1 and 2 clearly show that the convergence zone exists inside the ranges $-1.30 \leq \hbar_f \leq -0.15$, $-1.45 \leq \hbar_\theta \leq -0.25$ and $-1.40 \leq \hbar_\phi \leq -0.20$ for lower disk case $(\eta = 0)$ and $-1.10 \leq \hbar_f \leq -0.15$, $-1.35 \leq \hbar_\theta \leq -0.25$ and $-1.35 \leq \hbar_\phi \leq -0.10$ for upper disk case $(\eta = 1)$. Table 1 depicts that 16th order of deformations is sufficient for convergent homotopic solutions for lower disk, whereas the 18th order of deformations is necessary for convergent homotopic solutions regarding upper disks (see Table 2).

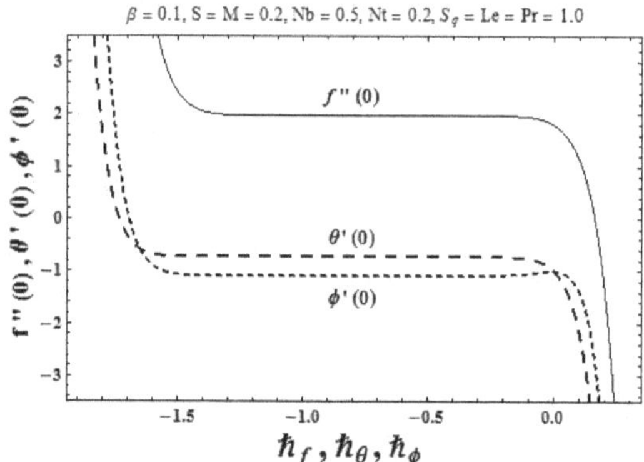

Figure 1. $\hbar$ -Curves for f, θ and ϕ at the lower disk.

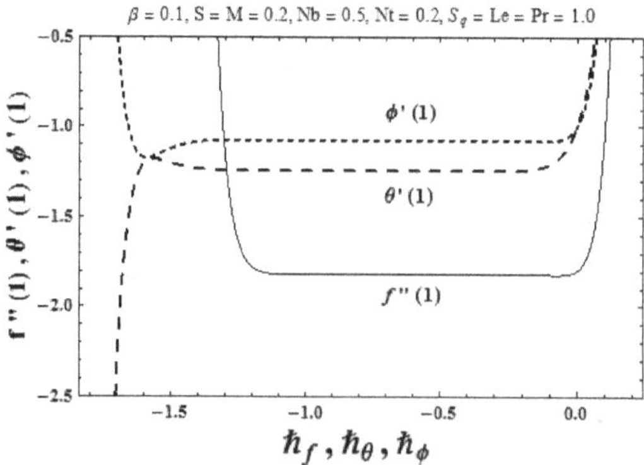

Figure 2. $\hbar$ -Curves for f, θ and ϕ at the upper disk.

Table 1. HAM solution convergence at the lower disk when $S = 0.2$, $M = Nt = 0.2$, $Nb = 0.5$ and $Sq = Le = Pr = 1.0$.

Order of Deformations	$f''(0)$	$-\theta'(0)$	$-\phi'(0)$
1	1.90121	0.08358	0.98916
5	1.98026	0.72074	1.07442
10	1.98313	0.71628	1.09068
16	1.98318	0.71623	1.09150
25	1.98318	0.71623	1.09150
35	1.98318	0.71623	1.09150
50	1.98318	0.71623	1.09150

Table 2. HAM solution convergence at the upper disk when $S = 0.2$, $M = Nt = 0.2$, $Nb = 0.5$ and $Sq = Le = Pr = 1.0$.

Order of Deformations	$-f''(1)$	$-\theta'(1)$	$-\phi'(1)$
1	1.82591	1.11083	1.06417
5	1.82053	1.23444	1.08069
10	1.81966	1.24300	1.07502
18	1.81965	1.24304	1.07475
25	1.81965	1.24304	1.07475
35	1.81965	1.24304	1.07475
50	1.81965	1.24304	1.07475

5. Discussion

This portion explores the effects of various pertinent parameters including Deborah number (β), Lewis number (Le), Brownian motion parameter (Nb), Prandtl number (Pr), thermophoresis parameter (Nt) and squeezing parameter (Sq) on temperature $\theta(\eta)$ and concentration $\phi(\eta)$ profiles. Figure 3 shows the the impact of Deborah number (β) on the temperature field $\theta(\eta)$. It is observed that the temperature field $\theta(\eta)$ decreases with the increase in the Deborah number (β). Figure 4 illustrates the impact of Brownian motion parameter (Nb) on temperature field $\theta(\eta)$. Here, temperature field $\theta(\eta)$ is increased by enhancing Brownian motion parameter. Variation of thermophoresis parameter (Nt) on temperature field $\theta(\eta)$ is sketched in Figure 5. Larger values of thermophoresis parameter (Nt) show higher temperature fields. Physically larger (Nt) causes an enhancement in temperature distribution. This is because of a stronger thermophoretic impact. Figure 6 shows temperature against Pr. Lower temperature is noticed for larger Pr. Figure 7 indicates that larger squeezing parameter (Sq) guarantees a decay in temperature $\theta(\eta)$. Figure 8 elucidates the impact of Deborah number (β) on the concentration profile $\phi(\eta)$. The concentration field $\phi(\eta)$ is decreased by increasing the Deborah number (β). Figure 9 shows the impact of Brownian motion parameter (Nb) on concentration profile $\phi(\eta)$. Concentration profile is reduced for larger values of Brownian motion parameter (Nb). Figure 10 shows behavior of thermophoresis parameter (Nt) on concentration field $\phi(\eta)$. Here, concentration field is enhanced for larger thermophoresis parameter. Figure 11 elucidates the concentration for variation of Lewis number (Le). Obviously larger (Le) leads to a large concentration field. Figure 12 sketched the concentration field $\phi(\eta)$ against Prandtl number (Pr). Larger (Pr) shows concentration field. Figure 13 declares that the increasing values of squeezing parameter (Sq) lead to higher enhancement. Table 3 is developed to validate the present results with the previously published results in a limiting sense. From this Table, we analyzed that the present HAM solution have good agreement with the previous solution by Hashmi et al. [6] in a limiting sense. Table 4 consists of skin friction at the lower and upper disks. Here, the skin friction coefficient at the lower and upper disks are higher for increasing Deborah number and squeezing parameter. Table 5 is computed to examine the numerical data of local Nusselt number at the lower and upper disks for several embedding parameters. It is observed that local Nusselt number enhances at both lower and upper disks for larger Lewis number while the reverse is found for Prandtl number. Table 6 depicts numerical data of local Sherwood number at the lower and upper disks for various values of pertinent parameters. Here, we noticed that local Sherwood number increases at both lower and upper disks for increasing values of squeezing parameter.

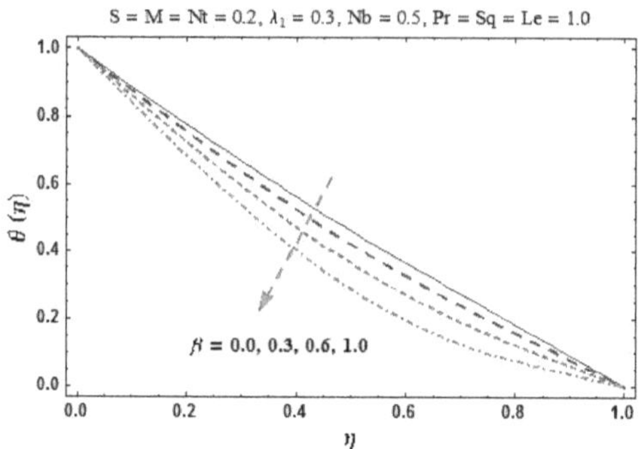

Figure 3. Plots of $\theta(\eta)$ for β.

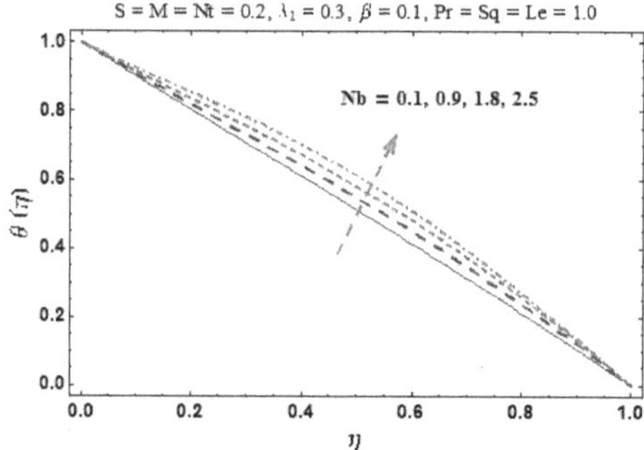

Figure 4. Plots of $\theta(\eta)$ for Nb.

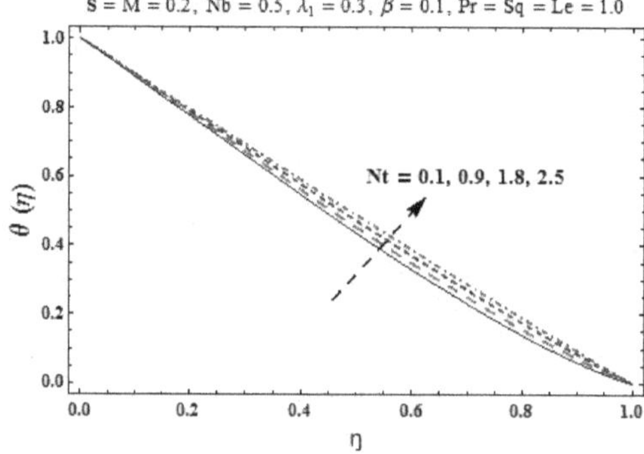

Figure 5. Plots of $\theta(\eta)$ for Nt.

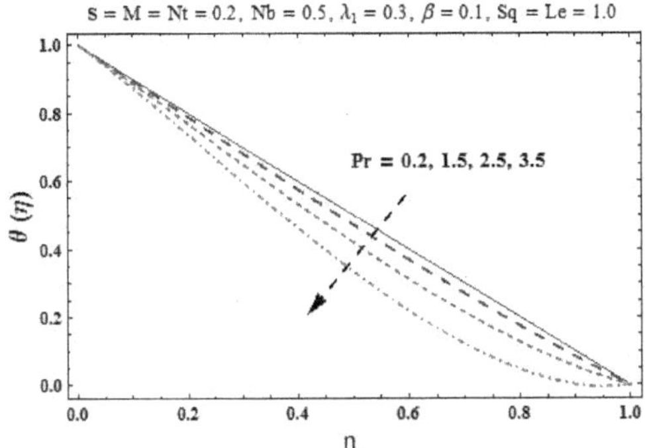

Figure 6. Plots of $\theta(\eta)$ for Pr.

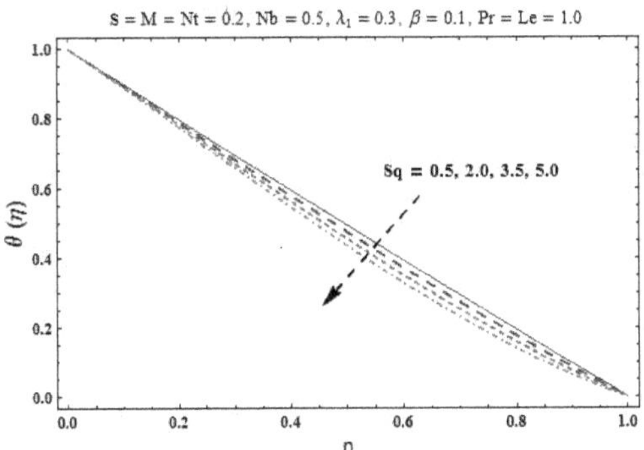

Figure 7. Plots of $\theta(\eta)$ for Sq.

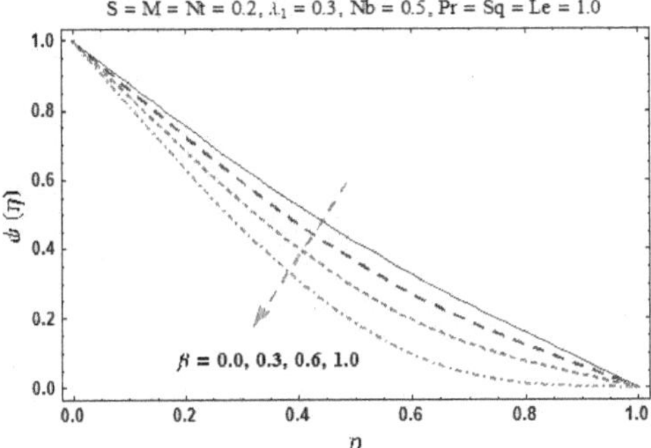

Figure 8. Plots of $\phi(\eta)$ for β.

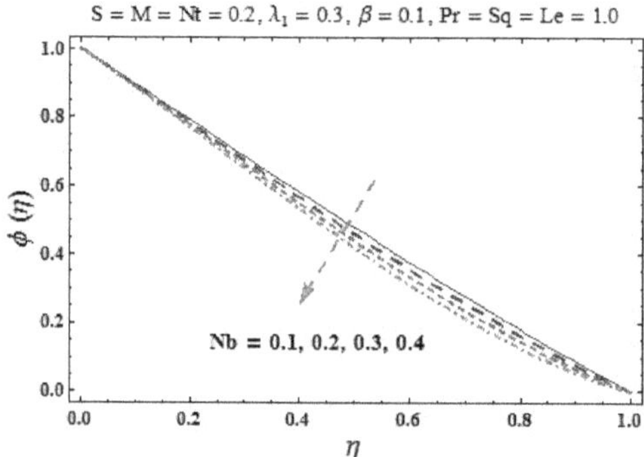

Figure 9. Plots of $\phi(\eta)$ for Nb.

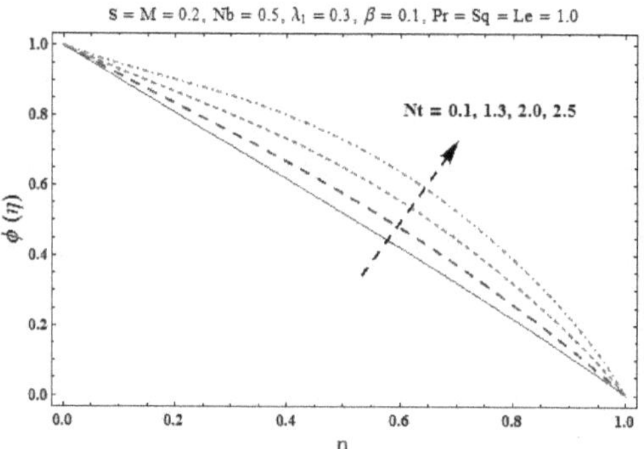

Figure 10. Plots of $\phi(\eta)$ for Nt.

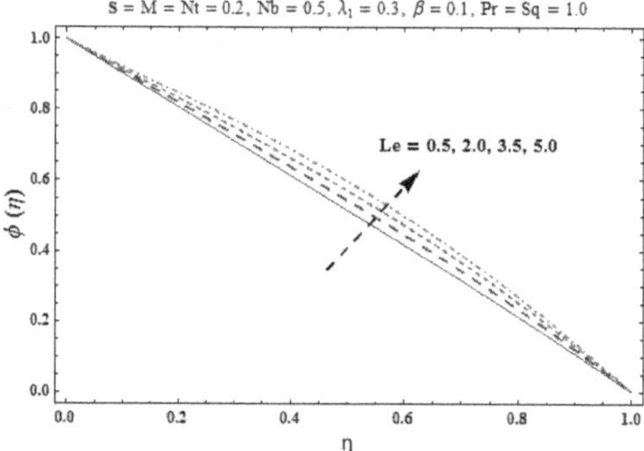

Figure 11. Plots of $\phi(\eta)$ for Le.

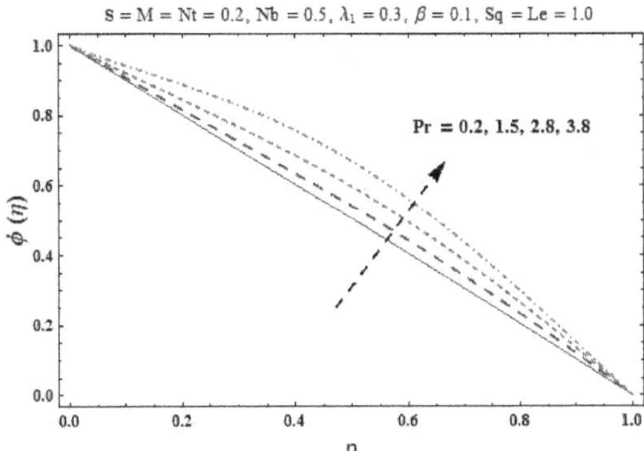

Figure 12. Plots of $\phi(\eta)$ for Pr.

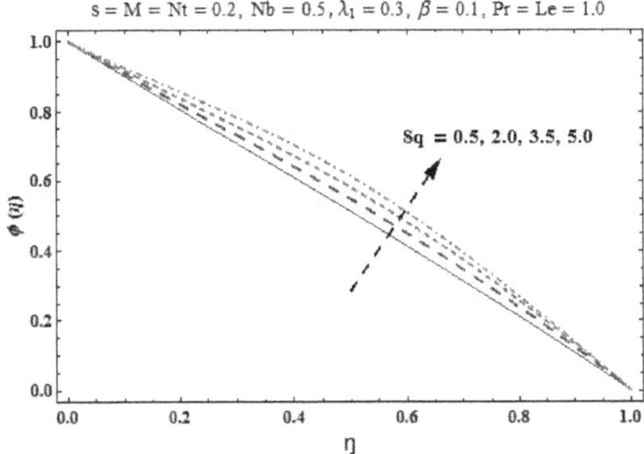

Figure 13. Plots of $\phi(\eta)$ for S_q.

Table 3. Comparative values of $f''(1)$ for different values of M when $Sq = 1.0$, $S = 2.0$ and $\beta = \lambda_1 = 0.0$.

M	$f''(1)$	
	HAM	**Hashmi et al. [6]**
0.0	7.533166	7.53316579
2.0	8.263872	8.26387230
3.0	9.097326	9.09732573
5.0	11.34929	11.3492890

Table 4. Skin friction coefficient at the lower and upper disks via S, β, λ_1, M and Sq.

S	β	λ_1	M	Sq	C_{f1}	C_{f2}
0.0	0.1	0.3	0.2	1.0	3.649479	3.590709
0.2	–	–	–	– –	2.287085	2.097308
0.4	–	–	–	–	0.795248	0.681540
0.2	0.0	0.3	0.2	1.0	2.002700	1.851550
–	0.1	–	–	–	2.280654	2.092601
–	0.2	–	–	–	2.554007	2.344925
0.2	0.1	0.0	0.2	1.0	2.232063	2.087393
–	–	0.1	–	–	2.248280	2.089086
–	–	0.2	–	–	2.264478	2.090822
0.2	0.1	0.3	0.0	1.0	2.279427	2.091703
–	–	–	0.5	–	2.287085	2.097308
–	–	–	1.0	–	2.309896	2.114041
0.2	0.1	0.3	0.2	0.5	2.175138	2.081827
–	–	–	–	1.0	2.280654	2.092601
–	–	–	–	1.5	2.385315	2.105140
–	–	–	–	2.0	2.489154	2.119203

Table 5. Numerical data for local Nusselt number at the lower and upper disks for several values of S, M, Sq, Nt, Nb, Le and Pr.

S	M	Sq	Nt	Nb	Le	Pr	$\theta'(0)$	$\theta'(1)$
0.0	0.2	1.0	0.2	0.5	1.0	1.0	0.77442	1.21503
0.5	–	–	–	–	–	–	0.63707	1.28291
1.0	–	–	–	–	–	–	0.52424	1.34412
0.5	0.0	1.0	0.2	0.5	1.0	1.0	0.63707	1.28291
–	0.5	–	–	–	–	–	0.63707	1.28291
–	1.0	–	–	–	–	–	0.63707	1.28291
0.5	0.2	0.0	0.2	0.5	1.0	1.0	0.69050	1.39050
–	–	1.0	–	–	–	–	0.63707	1.28291
–	–	2.0	–	–	–	–	0.58798	1.18403
0.5	0.2	1.0	0.0	0.5	1.0	1.0	0.71036	1.17119
–	–	–	0.5	–	–	–	0.53789	1.46215
–	–	–	1.0	–	–	–	0.39954	1.79063
0.5	0.2	1.0	0.2	0.5	1.0	1.0	0.71623	1.24304
–	–	–	–	1.0	–	–	0.54547	1.56082
–	–	–	–	1.5	–	–	0.40812	1.92538
0.5	0.2	1.0	0.2	0.5	0.5	1.0	0.63610	1.28096
–	–	–	–	–	1.0	–	0.63707	1.28291
–	–	–	–	–	1.5	–	0.63816	1.28511
0.5	0.2	1.0	0.2	0.5	1.0	0.5	0.80131	1.13712
–	–	–	–	–	–	1.0	0.63707	1.28291
–	–	–	–	–	–	1.5	0.50384	1.43981

Table 6. Numerical data for local Sherwood number at the lower and upper disks for several values of S, M, Sq, Nt, Nb, Le and Pr.

S	M	Sq	Nt	Nb	Le	Pr	$\phi'(0)$	$\phi'(1)$
0.0	0.2	1.0	0.2	0.5	1.0	1.0	0.99838	1.10381
0.5	–	–	–	–	–	–	1.23601	1.03275
1.0	–	–	–	–	–	–	1.49070	0.96658
0.5	0.0	1.0	0.2	0.5	1.0	1.0	1.23601	1.03275
–	0.5	–	–	–	–	–	1.23601	1.03275
–	1.0	–	–	–	–	–	1.23601	1.03275
0.5	0.2	0.0	0.2	0.5	1.0	1.0	1.12380	0.84379
–	–	1.0	–	–	–	–	1.23601	1.03275
–	–	2.0	–	–	–	–	1.33266	1.19929
0.5	0.2	1.0	0.0	0.5	1.0	1.0	1.07964	1.11899
–	–	–	0.5	–	–	–	1.58185	0.73579
–	–	–	1.0	–	–	–	2.39964	0.26741
0.5	0.2	1.0	0.2	0.5	1.0	1.0	1.09150	1.07475
–	–	–	–	1.0	–	–	1.07083	1.10001
–	–	–	–	1.5	–	–	1.06200	1.12874
0.5	0.2	1.0	0.2	0.5	0.5	1.0	1.19159	0.96141
–	–	–	–	–	1.0	–	1.23601	1.03275
–	–	–	–	–	1.5	–	1.27885	1.10206
0.5	0.2	1.0	0.2	0.5	1.0	0.5	1.12319	1.00307
–	–	–	–	–	–	1.0	1.23601	1.03275
–	–	–	–	–	–	1.5	1.33897	1.08419

6. Conclusions

Magnetohydrodynamic (MHD) squeezing flow of Jeffrey nanofluid between two parallel disks is examined. The key points of presented analysis are mentioned below:

- Larger values of Deborah number correspond to lower temperature and concentration profiles.
- Both temperature and concentration profiles are higher for larger values of thermophoresis parameter.
- Effects of Brownian motion parameter on temperature and concentration profiles are quite the opposite from each other.
- Larger values of Prandtl number show opposite trends for temperature and concentration profiles.
- Effects of squeezing parameter on temperature and concentration profiles are quite opposite to each other.
- The present analysis reduces to a Newtonian nanofluid flow situation when $\beta = \lambda_1 = 0$.

Acknowledgments: The authors are grateful for the useful suggestions of the reviewers.

Author Contributions: All authors contributed. Tasawar Hayat, Muhammad Ayub and Ahmed Alsaedi contributed in Sections 1, 2 and 5. Tehseen Abbas and Taseer Muhammad mainly have contribution in Sections 3, 4 and 5.

Conflicts of Interest: The authors declare no conflict of interest.

References

1. Choi, S.U.S. *Enhancing Thermal Conductivity of Fluids with Nanoparticles*; FED 231/MD; ASME: New York, NY, USA, 1995; pp. 99–105.
2. Buongiorno, J. Convective transport in nanofluids. *ASME J. Heat Transf.* **2006**, *128*, 240–250.
3. Makinde, O.D.; Aziz, A. Boundary layer flow of a nanofluid past a stretching sheet with a convective boundary condition. *Int. J. Therm. Sci.* **2011**, *50*, 1326–1332.
4. Mustafa, M.; Hayat, T.; Pop, I.; Asghar, S.; Obaidat, S. Stagnation-point flow of a nanofluid towards a stretching sheet. *Int. J. Heat Mass Transf.* **2011**, *54*, 5588–5594.

5. Turkyilmazoglu, M. Exact analytical solutions for heat and mass transfer of MHD slip flow in nanofluids. *Chem. Eng. Sci.* **2012**, *84*, 182–187.
6. Hashmi, M.M.; Hayat, T.; Alsaedi, A. On the analytic solutions for squeezing flow of nanofluid between parallel disks. *Nonlinear Anal. Model. Control* **2012**, *17*, 418–430.
7. Ibrahim, W.; Makinde, O.D. The effect of double stratification on boundary layer flow and heat transfer of nanofluid over a vertical plate. *Comput. Fluids* **2013**, *86*, 433–441.
8. Sheikholeslami, M.; Bandpy, M.G.; Ellahi, R.; Hassan, M.; Soleimani, S. Effects of MHD on Cu-water nanofluid flow and heat transfer by means of CVFEM. *J. Magn. Magn. Mater.* **2014**, *349*, 188–200.
9. Malvandi, A.; Safaei, M.R.; Kaffash, M.H.; Ganji, D.D. MHD mixed convection in a vertical annulus filled with Al_2O_3—Water nanofluid considering nanoparticle migration. *J. Magn. Magn. Mater.* **2015**, *382*, 296–306.
10. Hayat, T.; Muhammad, T.; Alsaedi, A.; Alhuthali, M.S. Magnetohydrodynamic three-dimensional flow of viscoelastic nanofluid in the presence of nonlinear thermal radiation. *J. Magn. Magn. Mater.* **2015**, *385*, 222–229.
11. Chamkha, A.; Abbasbandy, S.; Rashad, A.M. Non-Darcy natural convection flow for non-Newtonian nanofluid over cone saturated in porous medium with uniform heat and volume fraction fluxes. *Int. J. Numer. Methods Heat Fluid Flow* **2015**, *25*, 422–437.
12. Gireesha, B.J.; Gorla, R.S.R.; Mahanthesh, B. Effect of suspended nanoparticles on three-dimensional MHD flow, heat and mass transfer of radiating Eyring-Powell fluid over a stretching sheet. *J. Nanofluids* **2015**, *4*, 474–484.
13. Lin, Y.; Zheng, L.; Zhang, X.; Ma, L.; Chen, G. MHD pseudo-plastic nanofluid unsteady flow and heat transfer in a finite thin film over stretching surface with internal heat generation. *Int. J. Heat Mass Transf.* **2015**, *84*, 903–911.
14. Sheikholeslami, M.; Ellahi, R. Three dimensional mesoscopic simulation of magnetic field effect on natural convection of nanofluid. *Int. J. Heat Mass Transf.* **2015**, *89*, 799–808.
15. Hsiao, K.L. Stagnation electrical MHD nanofluid mixed convection with slip boundary on a stretching sheet. *Appl. Therm. Eng.* **2016**, *98*, 850–861.
16. Hayat, T.; Muhammad, T.; Shehzad, S.A.; Alsaedi, A. On three-dimensional boundary layer flow of Sisko nanofluid with magnetic field effects. *Adv. Powder Tech.* **2016**, *27*, 504–512.
17. Malvandi, A.; Ganji, D.D.; Pop, I. Laminar filmwise condensation of nanofluids over a vertical plate considering nanoparticles migration. *Appl. Therm. Eng.* **2016**, *100*, 979–986.
18. Hayat, T.; Imtiaz, M.; Alsaedi, A. Unsteady flow of nanofluid with double stratification and magnetohydrodynamics. *Int. J. Heat Mass Transf.* **2016**, *92*, 100–109.
19. Hayat, T.; Waqas, M.; Shehzad, S.A.; Alsaedi, A. On model of Burgers fluid subject to magneto nanoparticles and convective conditions. *J. Mol. Liq.* **2016**, *222*, 181–187.
20. Hayat, T.; Aziz, A. ; Muhammad, T.; Alsaedi, A. On magnetohydrodynamic three-dimensional flow of nanofluid over a convectively heated nonlinear stretching surface. *Int. J. Heat Mass Transf.* **2016**, *100*, 566–572.
21. Hussain, A.; Mohyud-Din, S.T.; Cheema, T.A. Analytical and numerical approaches to squeezing flow and heat transfer between two parallel disks with velocity slip and temperature jump. *Chin. Phys. Lett.* **2012**, *29*, 114705.
22. Chatraei, S.H.; Macosko, C.W.; Winter, H.H. Lubricated squeezing flow: A new biaxial extensional rheometer. *J. Rheol.* **1981**, *25*, 433–443.
23. Stefan, M.J. Versuch Uber die scheinbare adhesion, Sitzungsberichte der Akademie der Wissenschaften in Wien. *Math. Naturwissen* **1874**, *69*, 713–721.
24. Domairry, G.; Aziz, A. Approximate analysis of MHD squeeze flow between two parallel disks with suction or injection by homotopy perturbation method. *Math. Prob. Eng.* **2009**, *2009*, 603916.
25. Siddiqui, A.M.; Irum, S.; Ansari, A.R. Unsteady squeezing flow of a viscous MHD fluid between parallel plates, a solution using the homotopy perturbation method. *Math. Model. Anal.* **2008**, *13*, 565–576.
26. Rashidi, M.M.; Siddiqui, A.M.; Asadi, M. Application of homotopy analysis method to the unsteady squeezing flow of a second-grade fluid between circular plates. *Math. Probl. Eng.* **2010**, *2010*, 706840.
27. Qayyum, A.; Awais, M.; Alsaedi, A.; Hayat, T. Unsteady squeezing flow of Jeffery fluid between two parallel disks. *Chin. Phys. Lett.* **2012**, *29*, 034701.

28. Sheikholeslami, M.; Ganji, D.D. Heat transfer of Cu-water nanofluid flow between parallel plates. *Powder Technol.* **2013**, *235*, 873–879.
29. Hayat, T.; Muhammad, T.; Qayyum, A.; Alsaedi, A.; Mustafa, M. On squeezing flow of nanofluid in the presence of magnetic field effects. *J. Mol. Liq.* **2016**, *213*, 179–185.
30. Hayat, T.; Mustafa, M.; Shehzad, S.A.; Obaidat, S. Melting heat transfer in the stagnation-point flow of an upper-convected Maxwell (UCM) fluid past a stretching sheet. *Int. J. Numer. Methods Fluids* **2012**, *68*, 233–243.
31. Liao, S.J. On the homotopy analysis method for nonlinear problems. *Appl. Math. Comput.* **2004**, *147*, 499–513.
32. Dehghan, M.; Manafian, J.; Saadatmandi, A. Solving nonlinear fractional partial differential equations using the homotopy analysis method. *Numer. Methods Part. Differ. Equ.* **2010**, *26*, 448–479.
33. Malvandi, A.; Hedayati, F.; Domairry, G. Stagnation point flow of a nanofluid toward an exponentially stretching sheet with nonuniform heat generation/absorption. *J. Thermodyn.* **2013**, *2013*, 764827.
34. Abbasbandy, S.; Hashemi, M.S.; Hashim, I. On convergence of homotopy analysis method and its application to fractional integro-differential equations. *Quaest. Math.* **2013**, *36*, 93–105.
35. Arqub, O.A.; El-Ajou, A. Solution of the fractional epidemic model by homotopy analysis method. *J. King Saud Univ. Sci.* **2013**, *25*, 73–81.
36. Ellahi, R.; Hassan, M.; Zeeshan, A. Shape effects of nanosize particles in Cu-H_2O nanofluid on entropy generation. *Int. J. Heat Mass Transf.* **2015**, *81*, 449–456.
37. Hayat, T.; Imtiaz, M.; Alsaedi, A. Impact of magnetohydrodynamics in bidirectional flow of nanofluid subject to second order slip velocity and homogeneous—Heterogeneous reactions. *J. Magn. Magn. Mater.* **2015**, *395*, 294–302.
38. Sui, J.; Zheng, L.; Zhang, X.; Chen, G. Mixed convection heat transfer in power law fluids over a moving conveyor along an inclined plate. *Int. J. Heat Mass Transf.* **2015**, *85*, 1023–1033.
39. Hayat, T.; Hussain, Z.; Muhammad, T.; Alsaedi, A. Effects of homogeneous and heterogeneous reactions in flow of nanofluids over a nonlinear stretching surface with variable surface thickness. *J. Mol. Liq.* **2016**, *221*, 1121–1127.
40. Hayat, T.; Abbas, T.; Ayub, M.; Farooq, M.; Alsaedi, A. Flow of nanofluid due to convectively heated Riga plate with variable thickness. *J. Mol. Liq.* **2016**, *222*, 854–862.
41. Hsiao, K.L. MHD mixed convection for viscoelastic fluid past a porous wedge. *Int. J. Non-Linear Mech.* **2011**, *46*, 1–8.
42. Hsiao, K.L. Combined electrical MHD heat transfer thermal extrusion system using Maxwell fluid with radiative and viscous dissipation effects. *Appl. Therm. Eng.* **2016**, *8*, 208.
43. Hsiao, K.L. Numerical solution for Ohmic Soret-Dufour heat and mass mixed convection of viscoelastic fluid over a stretching sheet with multimedia physical features. *J. Aerosp. Eng.* **2016**, doi:10.1061/(ASCE)AS.1943-5525.0000681.

Article

Thin Film Williamson Nanofluid Flow with Varying Viscosity and Thermal Conductivity on a Time-Dependent Stretching Sheet

Waris Khan [1], Taza Gul [2,*], Muhammad Idrees [1], Saeed Islam [3], Ilyas Khan [4] and L.C.C. Dennis [5,*]

1 Departement of Mathematics, Islamia College Peshawar, Peshawar 25000, Pakistan; wariskhan758@yahoo.com (W.K.); idreesgiki@gmail.com (M.I.); Tel.: +92-302-839-7271 (W.K.); +92-332-900-2556 (M.I.)
2 Higher Education Department Khyber Pakhtunkhwa, Peshawar 25000, Pakistan
3 Departement of Mathematics, Abdul Wali Khan University, Mardan 32300, Pakistan; saeedislam@awkum.edu.pk; Tel.: +92-334-585-5248
4 Department of Mathematics, College of Engineering, Majmaah University, Majmaah 31750, Saudi Arabia; ilyaskhanqau@yahoo.com; Tel.: +966-503-346-170
5 Department of Fundamental and Applied Sciences, Universiti Teknologi Petronas, Perak 32610, Malaysia
* Correspondence: tazagulsafil@yahoo.com (T.G.); dennis.ling@petronas.com.my (L.C.C.D.); Tel.: +92-331-911-7160 (T.G.); +60-168-529-529-339 (L.C.C.D.)

Academic Editor: Rahmat Ellahi
Received: 17 August 2016; Accepted: 21 October 2016; Published: 15 November 2016

Abstract: This article describes the effect of thermal radiation on the thin film nanofluid flow of a Williamson fluid over an unsteady stretching surface with variable fluid properties. The basic governing equations of continuity, momentum, energy, and concentration are incorporated. The effect of thermal radiation and viscous dissipation terms are included in the energy equation. The energy and concentration fields are also coupled with the effect of Dufour and Soret. The transformations are used to reduce the unsteady equations of velocity, temperature and concentration in the set of nonlinear differential equations and these equations are tackled through the Homotopy Analysis Method (HAM). For the sake of comparison, numerical (ND-Solve Method) solutions are also obtained. Special attention has been given to the variable fluid properties' effects on the flow of a Williamson nanofluid. Finally, the effect of non-dimensional physical parameters like thermal conductivity, Schmidt number, Williamson parameter, Brinkman number, radiation parameter, and Prandtl number has been thoroughly demonstrated and discussed.

Keywords: Williamson fluid; unsteady flow; nanofluid film; HAM and numerical method

1. Introduction

The fluid flow on a nonlinear stretching surface has attracted the attention of several investigators due to its numerous applications in the fields of engineering and industry, such as oil filtering processes, paper making processes, polymer making, food manufacturing and preserving processes, etc. The flow provides more effective results in the manufacturing of good quality products in the engineering field when heat is transferred to it, for instance via metallurgical processes, wire and fiber coating, heat exchange equipment, the polymers extrusion process, the chemical polymer process, good quality glass manufacturing and crystal growing, and so on. In case of a slow cooling rate and stretching rate of electrically conducted fluids, magneto hydrodynamic (MHD) flow provides the best quality products [1]. Sakiadis [2] was the pioneer to study the flow on a linearly stretched surface when the fluid was at rest. Crane [3] examined the flow on the stretching sheet and obtained a similar solution to the problem. He also obtained a closed form exponential solution to the linear flow on the stretching sheet. The suction and blowing process together with heat and mass transmission rate over the stretched sheet were formulated by Gupta and Gupta [4]. Elbashbeshy [5] inspected the flow on the stretched surface with inconstant heat flux. Aziz [6] investigated the flow on an unsteady stretching sheet and observed the heat radiation effect. Mukhopadyay [7] later considered thermal radiation's effect on a vertically stretched surface with a porous medium. Shateyi and Motsa [8] discussed heat and mass transfer rates over a horizontal stretched surface numerically. Aziz [9] investigated momentum and the heat effect on an electric current providing and incompressible fluid over a linear stretching surface. Hady et al. [10] extended the abovementioned work and discussed heat transfer and radiation effect on viscous flow of a nanofluid over a non-linearly stretched surface. Pavlov [11] examined the MHD flow of a viscous fluid with constant density over a linear stretched surface. Bianco et al. [12] investigated the second principle of thermodynamics applied to a water–Al_2O_3 nanofluid. They studied that how the generation of entropy within the tube varies if inlet conditions, particle concentrations, and dimensions are changed. Nadeem et al. [13–16] investigated a variety of fluid models on the stretching surface by taking linear as well as exponential sheets. Such flow nowadays has many applications in the fields of physics, chemistry, and engineering; processes such as the cooling of an electro-magnetic fluid on a stretching sheet can be used to make a good quality thinning copper wire. Suction and blowing processes, and heat and mass transferring with time-dependent surface, were analyzed by Elbashbeshy and Bazid [17].

The viscosity effect and thermal conductivity behavior of the fluid are taken as constant in all of the studies discussed above. The physical properties of a fluid strongly depend on the temperature. Experimentally, it has been proven that the magnitude of viscosity is directly related to the temperature of gases and inversely proportional in the case of liquids. However, the thermal conductivity property of the fluid is directly proportional to the temperature. Variable viscosity, thermal conductivity, or a combination of these two are studied in several research articles; for instance, Grubka and Bobba [18] measured the flow on a horizontally moving stretched sheet while the temperature of the surface was considered variable. Chen and Char [19] obtained the particular solution for the variable heat flux on a surface when force was applied. Pop et al. [20] and Pantokratoras [21] investigated varying viscosity's and heat transfer's effect, respectively, on moving plates. It is also shown that the effect of temperature is inversely proportional to fluid viscosity. Abel et al. [22] investigated the flow of visco-elastic fluids on a porous stretching surface with variable fluid viscosity. The temperature function is inversely related to fluid viscosity and a fourth-order RK method was used to solve the combined nonlinear equations. Makinde and Mishra [23] investigated the combined effects of variable viscosity, Brownian motion, and thermophoresis in the water base nanofluids past a stretching surface. They used a shooting method for the solution of coupled differential equations and discussed the effect of flow parameters. Mukhopadhyay et al. [24] examined the MHD effects of heated fluids of variable viscosity on a stretched surface. It is also assumed that fluid viscosity is related linearly to temperature. The equations related to flow pattern are simplified by using scaling group transformations and then a numerical method was used to solve the resulting non-linear ordinary differential equations. Fourier's Law illustrates the association between energy fluctuation and the gradient of temperature, while Fick's Law shows the association between the mass fluctuation and concentration gradient. However, in 1873, Dufour showed that the energy fluctuation is also affected by configuration gradient, so it was named the Dufour effect or the diffusion-thermo effect. Soret observed that mass fluctuation is created by temperature gradient, so it is called the thermal diffusion effect. This effect is very important in the flow when there is a density difference. Hayat et al. [25] examined the Soret and Dufour effects over an exponential stretching surface with a spongy medium. Alam et al. [26] examined the 2D free convection flow over the semi-infinite perpendicular porous surface containing the effects of Soret and Dufour numbers. Kafoussias and Williams [27] studied the mixed convection flow and considered the heat and mass transmission, keeping the temperature flux variable and observing the Soret and Dufour effects, respectively. Chamkha and Ben-Nakhi [28] considered the mixed convection pattern flow over the perpendicular permeable porous surface in view of the effects of magnetic and thermal radioactivity and discussed the Soret effect and Dufour effect. The effects of Soret number and Dufour number on free convective flow over a stretched surface were investigated by Afify [29] with heat and mass transmission. Beg et al. [30] considered the effect of Soret and Dufour numbers over a free-convective saturated spongy surface in the presence of MHD heat and mass transmission. El-Kabeir et al. [31] investigated the effects of Dufour and Soret numbers over a non-Darcy spherically porous natural convection MHD heat and mass transmission. The special effects of Soret number and Dufour number of non-Darcy instable mixed convective MHD flow over the stretched medium, considering heat and mass transmission, were investigated by Pal and Mondal [32]. Yasir et al. [33] analyzed the effects of variable viscosity and thermal conductivity on a thin film flow over a shrinking/stretching sheet. Aziz et al. [34] investigated thin film flow and heat transfer on an unsteady stretching sheet with internal heating. Qasim et al. [35] discussed heat and mass transfer in a nanofluid over an unsteady stretching sheet using Buongiorno's model. Prashant et al. [36] analyzed thin film flow and heat transfer on an unsteady stretching sheet with thermal radiation and internal heating in presence of external magnetic field. The published work is incomplete, though for both of these physical parameter there exist numerous industrial and mechanical applications. The few other investigations in this direction were made by Ellahi et al. [37], Akbar et al. [38,39], Shehzad et al. [40], and Zeeshan et al. [41].

The current work is the study of thin film flow of a Williamson nanofluid with the combined effect of varying thermal conductivity and viscosity on a time-dependent stretching sheet. The effect of Dufour and Soret numbers is discussed in detail. Also, the effects of Schmidt number and Brinkman number, thermal contamination, and viscous dissipation are considered. Applying these suppositions and similarity transformation on the governing partial differential equations (PDEs) of the flow is converted to non-linear ordinary differential equations (ODEs) and then solved through HAM [42–48]. The related work to the given flow is also discussed in [49–51].

The literature survey shows that there have been several investigations on nanofluids. However, so far, no study has been reported about the analysis of thin film flows of a Williamson nanofluid flow in two dimensions. The present study aims to analyze the variable thermal conductivity and viscosity of a two-dimensional thin film Williamson nanofluid past a stretching sheet.

2. Materials and Methods

Consider a two-dimensional flow of Williamson fluid that has constant density, variable viscosity, and a temperature gradient over an unsteady stretched surface, in which heat and mass are transmitted instantaneously. The flow coordinates are selected in such a manner that the x-axis is parallel to the plate and the y-axis is vertical to it. The stretching velocity of the sheet is in the direction of the x-axis with magnitude $U(x,t) = \frac{bx}{1-at}$, in which $b > 0$ is the stretching velocity constraint and defined in [37–39]. If $b < 0$ then it will become a shrinking velocity constraint. The temperature field is defined as $T_s(x,t) = T_0 - T_{ref}\left[\frac{bx^2}{2\upsilon}\right](1-at)^{-\frac{3}{2}}$, and the magnitude is inversely proportional to the distance from the surface. Similarly, the concentration field for the given flow is defined as $C_s(x,t) = C_0 - C_{ref}\left[\frac{bx^2}{2\upsilon}\right](1-at)^{-\frac{3}{2}}$, where T_0 represents the temperature at the surface, T_{ref} indicates the reference temperature, and C_{ref} indicates the reference concentration, respectively, as shown in [27–30], such that $0 \leq T_{ref} \leq T_0$ and $0 \leq C_{ref} \leq C_0$. The local Reynolds is defined as $\frac{bx^2}{\upsilon(1-at)}$. Firstly, the sheet is fixed to the origin; after that some outer force is applied to stretch the surface in the direction of the x-coordinate axis at a velocity $U(x,t) = \frac{b}{1-at}$ in time $0 \leq a < 1$.

Taking the above suppositions into consideration, the governing equations of continuity, velocity, temperature, and concentration can be expressed as:

$$\frac{\partial u}{\partial x} + \frac{\partial v}{\partial y} = 0, \tag{1}$$

$$\frac{\partial u}{\partial t} + u\frac{\partial u}{\partial x} + v\frac{\partial u}{\partial y} = \frac{1}{\rho}\frac{\partial}{\partial y}\left(\mu(T)\frac{\partial u}{\partial y}\right) + \frac{\sqrt{2}\Gamma}{\rho}\frac{\partial}{\partial y}\left[\mu(T)\frac{\partial u}{\partial y}\right]\frac{\partial u}{\partial y}, \tag{2}$$

$$\rho c_p\left[\frac{\partial T}{\partial t}+u\frac{\partial T}{\partial x}+v\frac{\partial T}{\partial y}\right]=\frac{\partial}{\partial y}\left[k(T)\frac{\partial T}{\partial y}\right]-\frac{\partial q_r}{\partial y}+\mu(T)\left[\left(\frac{\partial u}{\partial y}\right)^2+\sqrt{2}\Gamma\left(\frac{\partial u}{\partial y}\right)^3\right], \\ +\frac{\rho D_m k_T}{c_s}\frac{\partial^2 C}{\partial y^2} \tag{3}$$

$$\left[\frac{\partial C}{\partial t}+u\frac{\partial C}{\partial x}+v\frac{\partial C}{\partial y}\right]=D_m\frac{\partial^2 C}{\partial y^2}+\frac{D_m k_T}{T_m}\frac{\partial^2 T}{\partial y^2}. \tag{4}$$

The boundary conditions are:

$$u=U,\ v=0,\ T=T_s,\ C=C_s, \tag{5}$$

$$\frac{\partial u}{\partial y}=\frac{\partial T}{\partial y}=\frac{\partial C}{\partial y}=0\quad v=\frac{dh}{dt}=0, \tag{6}$$

where $\mu(T)=\frac{\mu_0}{(1-\gamma)\frac{T-T_0}{T_{ref}(\frac{bx^2}{2\nu})}}$ indicates the variable viscosity in which μ_0 is the fluid viscosity at reference temperature T_0 and the coefficient γ expresses the strength of the dependency between μ and T. $K(T)=K_1\left(1-\varepsilon\left(\frac{T-T_0}{T_{ref}(\frac{bx^2}{2\nu})}\right)\right)$ represents the temperature-dependent thermal conductivity, in which ε is the variable thermal conductivity parameter. The kinematics viscosity is represented as $\upsilon=\frac{\mu_0}{\rho}$, $\Gamma>0$ is the time constant, u and v are the velocities along the x-axis and y-axis, respectively, T and C represent the temperature and concentration fields, respectively, ρ indicates the density of the fluid, C_p designates the specific heat, C_s represents the absorption susceptibility, liquid film thickness is denoted by $h(t)$, $q_r=-\frac{16\sigma T_s^3}{3k}\frac{\partial T}{\partial y}$ indicates the radiative heat fluctuation, the Stefan–Boltzmann constant is specified by σ, the species concentration molecular diffusivity is represented by D_m, T_m represents the mean temperature, the thermal diffusion ratio is denoted by k_T, and k designates the thermal conductivity of the liquid film.

We introduced the following transformations for the velocity, temperature, and concentration fields:

$$\begin{aligned}&\psi(x,y,t)=x\sqrt{\tfrac{\upsilon b}{1-at}}f(\eta),u=\tfrac{\partial\psi}{\partial y}=\tfrac{bx}{(1-at)}f'(\eta)=\tfrac{\beta^2 x\nu}{h^2}f'(\eta),\\&v=-\tfrac{\partial\psi}{\partial x}=-\sqrt{\tfrac{\upsilon b}{1-at}}f(\xi)=-\tfrac{\nu\beta}{h}f(\eta),\eta=\sqrt{\tfrac{b}{\upsilon(1-at)}}y=\tfrac{\beta}{h}y,\\&T(x,y,t)=T_0-T_{ref}\left[\tfrac{bx^2}{2\upsilon}\right](1-at)^{-\frac{3}{2}}\theta(\eta),C(x,y,t)=C_0-C_{ref}\left[\tfrac{bx^2}{2\upsilon}\right](1-at)^{-\frac{3}{2}}\varphi(\eta),\end{aligned} \tag{7}$$

where a prime number specifies the derivative with respect to η and $\psi(x,y,t)$ is the stream function; $\beta=h(t)\sqrt{\frac{b}{\upsilon(1-\alpha t)}}$ is the non-dimensional thickness of the nano liquid film and $h(t)$ is the uniform thickness of the fluid film, which gives $\frac{dh}{dt}=-\frac{\beta a}{2}\left[\frac{\upsilon}{b}\right]^{\frac{1}{2}}(1-\alpha t)^{-\frac{1}{2}}$.

Plugging the similarity variables from Equation (7) into Equations (1)–(6) satisfies the continuity equation, and the leftover equations are converted to couple nonlinear differential equations:

$$f'''+\lambda f''f'''+(1+\Lambda\theta)\left[ff''-(f')^2-S\left(f'+\frac{\eta}{2}f''\right)\right]=0 \tag{8}$$

$$\begin{aligned}&(1+\varepsilon\theta+N_r)(1+\Lambda\theta)\theta''-Pr(1+\Lambda\theta)\left(\tfrac{S}{2}(3\theta+\eta\theta')-f\theta'+2f'\theta\right)+\\&B_r\left((f'')^2+\lambda(f'')^3\right)+Pr(1+\Lambda\theta)D_u\varphi''=0,\end{aligned} \tag{9}$$

$$\varphi''+S_cS_r\theta''-S_c\left(\frac{S}{2}(3\varphi+\eta\varphi')+2f'\varphi-f\varphi'\right)=0. \tag{10}$$

The boundary conditions are transformed to:

$$\begin{aligned}&f(0)=0,\ f\prime(0)=1,\ f(\beta)=\tfrac{S\beta}{2},\ f''(\beta)=0,\\&\theta(0)=\varphi(0)=1,\ \theta'(\beta)=\varphi'(\beta)=0.\end{aligned} \tag{11}$$

Here $\Lambda = \gamma(T_s - T_0)$ represents the variable viscosity parameter, $Pr = \frac{\rho\upsilon c_p}{k}$ is the Prandtl number, $S = \frac{a}{b}$ is the non-dimensional measure of unsteadiness, $D_u = \frac{D_m k_T}{\upsilon c_p c_s}\frac{(C_s - C_0)}{(T_s - T_0)}$ is the Dufour number, $S_c = \frac{\upsilon}{D_m}$ is used for the Schmidt number, $S_r = \frac{D_m k_T}{\upsilon T_m}\frac{(T_s - T_0)}{(C_s - C_0)}$ represents the Soret number, $B_r = \frac{\mu_0 U_0^2}{k(T_s - T_0)}$ is the Brinkman number $N_r = \frac{16 T_\infty^3 \sigma_1}{3kk^*}$ indicates the thermal radiation parameter, and $\lambda = \Gamma x\sqrt{\frac{2b^3}{\upsilon(1-at)^3}}$ is the Williamson fluid constant.

Solution by HAM

In order to solve Equations (8)–(10) under the boundary conditions (11), we use the Homotopy Analysis Method (HAM) with the following procedure. The solutions having the auxiliary parameters $\hbar$ regulate and control the convergence of the solutions.

The initial guesses are selected as follows:

$$f_0(\eta) = \eta,\ \theta_0(\eta) = 1 \text{ and } \varphi_0(\eta) = 1. \tag{12}$$

The linear operators are taken as L_f, L_θ and L_φ:

$$L_f(f) = f''',\ L_\theta(\theta) = \theta'' \text{ and } L_\varphi(\varphi) = \varphi'', \tag{13}$$

which have the following properties:

$$L_f(c_1 + c_2\eta + c_3\eta^2) = 0,\ L_\theta(c_4 + c_5\eta) = 0 \text{ and } L_\varphi(c_6 + c_7\eta) = 0, \tag{14}$$

where $c_i (i = 1-7)$ are the constants in general solution:

The resultant non-linear operatives N_f, N_θ and N_φ are given as:

$$\begin{aligned} N_f[f(\eta;p)] &= \frac{\partial^3 f(\eta;p)}{\partial\eta^3} + \lambda\frac{\partial^2 f(\eta;p)}{\partial\eta^2}\frac{\partial^3 f(\eta;p)}{\partial\eta^3} \\ &+ (1 + \Lambda\theta(\eta;p))\left[f(\eta;p)\frac{\partial^2 f(\eta;p)}{\partial\eta^2} - \left(\frac{\partial f(\eta;p)}{\partial\eta}\right)^2 - S\left(\frac{\partial f(\eta;p)}{\partial\eta} + \frac{\eta}{2}\frac{\partial^2 f(\eta;p)}{\partial\eta^2}\right)\right], \end{aligned} \tag{15}$$

$$\begin{aligned} &N_\theta[f(\eta;p),\theta(\eta;p),\varphi(\eta;p)] = (1 + \varepsilon\theta(\eta;p) + N_r)(1 + \Lambda\theta(\eta;p))\frac{\partial^2\theta(\eta;p)}{\partial\eta^2} - \\ &\Pr(1 + \Lambda\theta(\eta;p))\left[\frac{S}{2}\left(3\theta(\eta;p) + \eta\frac{\partial\theta(\eta;p)}{\partial\eta}\right) + 2\theta(\eta;p)\frac{\partial f(\eta;p)}{\partial\eta} - f(\eta;p)\frac{\partial\theta(\eta;p)}{\partial\eta}\right] \\ &+B_r\left[\left(\frac{\partial^2 f(\eta;p)}{\partial\eta^2}\right)^2 + \lambda\left(\frac{\partial^2\theta(\eta;p)}{\partial\eta^2}\right)^3\right] + \Pr D_u(1 + \Lambda\theta(\eta;p))\frac{\partial^2\varphi(\eta;p)}{\partial\eta^2}, \end{aligned} \tag{16}$$

$$\begin{aligned} &N_\varphi[f(\eta;p),\theta(\eta;p),\varphi(\eta;p)] = \frac{\partial^2\varphi(\eta;p)}{\partial\eta^2} + S_c S_r\frac{\partial^2\theta(\eta;p)}{\partial\eta^2} - \\ &S_c\left[\frac{S}{2}\left(3\varphi(\eta;p) + \eta\frac{\partial\varphi(\eta;p)}{\partial\eta}\right) + 2\varphi(\eta;p)\frac{\partial f(\eta;p)}{\partial\eta} - f(\eta;p)\frac{\partial\varphi(\eta;p)}{\partial\eta}\right]. \end{aligned} \tag{17}$$

The basic idea of HAM is described in [32–35]; the zero-order problems from Equations (8)–(10) are:

$$(1-p)L_f[f(\eta;p) - f_0(\eta)] = p\hbar_f N_f[f(\eta;p)] \tag{18}$$

$$(1-p)L_\theta[\theta(\eta;p) - \theta_0(\eta)] = p\hbar_\theta N_\theta[f(\eta;p),\theta(\eta;p),\varphi(\eta;p)] \tag{19}$$

$$(1-p)L_\varphi[\varphi(\eta;p) - \varphi_0(\eta)] = p\hbar_\varphi N_\varphi[f(\eta;p),\theta(\eta;p),\varphi(\eta;p)]. \tag{20}$$

The equivalent boundary conditions are:

$$\begin{aligned} &f(\eta;p)|_{\eta=0} = 0,\ \left.\frac{\partial f(\eta;p)}{\partial\eta}\right|_{\eta=0} = 1,\ \left.\frac{\partial^2 f(\eta;p)}{\partial\eta^2}\right|_{\eta=\beta} = 0, \\ &\theta(\eta;p)|_{\eta=0} = 1,\ \left.\frac{\partial\theta(\eta;p)}{\partial\eta}\right|_{\eta=\beta} = 0,\ \varphi(\eta;p)|_{\eta=0} = 1,\ \left.\frac{\partial\varphi(\eta;p)}{\partial\eta}\right|_{\eta=\beta} = 0\ , \end{aligned} \tag{21}$$

where $p \in [0,1]$ is the imbedding parameter, and $\hbar_f$, $\hbar_\theta$ and $\hbar_\varphi$ are used to control the convergence of the solution. When $p = 0$ and $p = 1$ we have:

$$f(\eta;1) = f(\eta),\ \theta(\eta;1) = \theta(\eta) \text{ and } \varphi(\eta;1) = \varphi(\eta). \tag{22}$$

Expanding $f(\eta;p)$, $\theta(\eta;p)$ and $\varphi(\eta;p)$ in Taylor's series about $p = 0$, we get

$$\begin{aligned} f(\eta;p) &= f_0(\eta) + \sum_{m=1}^{\infty} f_m(\eta)p^m, \\ \theta(\eta;p) &= \theta_0(\eta) + \sum_{m=1}^{\infty} \theta_m(\eta)p^m, \quad , \\ \varphi(\eta;p) &= \varphi_0(\eta) + \sum_{m=1}^{\infty} \varphi_m(\eta)p^m. \end{aligned} \tag{23}$$

where

$$f_m(\eta) = \frac{1}{m!}\frac{\partial f(\eta;p)}{\partial \eta}\bigg|_{p=0},\ \theta_m(\eta) = \frac{1}{m!}\frac{\partial \theta(\eta;p)}{\partial \eta}\bigg|_{p=0} \text{ and } \varphi_m(\eta) = \frac{1}{m!}\frac{\partial \varphi(\eta;p)}{\partial \eta}\bigg|_{p=0}. \tag{24}$$

The secondary constraints $\hbar_f$, $\hbar_\theta$ and $\hbar_\varphi$ are chosen in such a way that the series in Equation (23) converges at $p = 1$, we obtain:

$$\begin{aligned} f(\eta) &= f_0(\eta) + \sum_{m=1}^{\infty} f_m(\eta), \\ \theta(\eta) &= \theta_0(\eta) + \sum_{m=1}^{\infty} \theta_m(\eta), \\ \varphi(\eta) &= \varphi_0(\eta) + \sum_{m=1}^{\infty} \varphi_m(\eta). \end{aligned} \tag{25}$$

The m th-order problem satisfies the following:

$$\begin{aligned} L_f\left[f_m(\eta) - \chi_m f_{m-1}(\eta)\right] &= \hbar_f R_m^f(\eta), \\ L_\theta\left[\theta_m(\eta) - \chi_m \theta_{m-1}(\eta)\right] &= \hbar_\theta R_m^\theta(\eta), \\ L_\varphi\left[\varphi_m(\eta) - \chi_m \varphi_{m-1}(\eta)\right] &= \hbar_\varphi R_m^\varphi(\eta). \end{aligned} \tag{26}$$

The corresponding boundary conditions are:

$$\begin{aligned} &f_m(0) = f'_m(0) = \theta_m(0) = \varphi_m(0) = 0, \\ &f''_m(\beta) = \theta'_m(\beta) = \varphi'_m(\beta) = 0. \end{aligned} \tag{27}$$

Here

$$\begin{aligned} R_m^f(\eta) = {} & f'''_{m-1} + \lambda\sum_{k=0}^{m-1} f''_{m-1-k}f'''_k + \left[f''_{m-1} - \sum_{k=0}^{m-1} f'_{m-1-k}f'_k - S\left(f'_{m-1} + \tfrac{\eta}{2}f''_{m-1}\right)\right] + \\ & \Lambda\left[\sum_{k=0}^{m-1} \theta_{m-1-k}f''_k - \sum_{k=0}^{m-1} \theta_{m-1-k}\sum_{l=0}^{k} f'_{k-l}f'_l - S\left(\sum_{k=0}^{m-1} \theta_{m-1-k}f'_k + \tfrac{\eta}{2}\sum_{k=0}^{m-1} \theta_{m-1-k}f''_k\right)\right], \end{aligned} \tag{28}$$

$$\begin{aligned} R_m^\theta(\eta) = {} & (1+N_r)\,\theta''_{m-1} + (\varepsilon + \Lambda(1+N_r))\sum_{k=0}^{m-1} \theta_{m-1-k}\theta'_k + \varepsilon\Lambda\sum_{k=0}^{m-1} \theta_{m-1-k}\sum_{l=0}^{k} \theta_{k-l}\theta''_l - \\ & \Pr\left[\tfrac{S}{2}\left(3\theta_{m-1} + \eta\theta'_{m-1}\right) + 2\sum_{k=0}^{m-1} \theta_{m-1-k}f'_k - \sum_{k=0}^{m-1} f_{m-1-k}\theta'_k\right] - \\ & \Lambda\Pr\left[\tfrac{S}{2}\left(3\sum_{k=0}^{m-1} \theta_{m-1-k}\theta_k + \eta\sum_{k=0}^{m-1} \theta_{m-1-k}\theta'_k\right) + 2\sum_{k=0}^{m-1} \theta_{m-1-k}\sum_{l=0}^{k} \theta_{k-l}f'_l - \sum_{k=0}^{m-1} \theta_{m-1-k}\sum_{l=0}^{k} f_{k-l}\theta'_l\right] + \\ & B_r\left[\sum_{k=0}^{m-1} f''_{m-1-k}f''_k + \lambda\sum_{k=0}^{m-1} f''_{m-1-k}\sum_{l=0}^{k} f''_{k-l}f''_l\right] + \Pr D_u(1+\Lambda)\left[\varphi''_{\omega-1} + \sum_{k=0}^{m-1} \theta_{m-1-k}\varphi''_k\right], \end{aligned} \tag{29}$$

$$R_m^\varphi(\eta) = \varphi''_{m-1} + S_rS_c\theta''_{m-1} - S_c\left[\tfrac{S}{2}\left(3\varphi_{m-1} + \eta\varphi'_{m-1}\right) + \sum_{k=0}^{m-1} f'_{m-1-k}\varphi_k - \sum_{k=0}^{m-1} f_{m-1-k}\varphi'_j\right], \tag{30}$$

where

$$\chi_m = \begin{cases} 0, \text{ if } p \leq 1 \\ 1, \text{ if } p > 1 \end{cases}$$

3. Results

The Figure 1 represent geometry of the problem. The convergence of the series given in Equation (25), $f(\eta)$, $\theta(\eta)$, and $\varphi(\eta)$ entirely depend upon the auxiliary parameters $\hbar_f$, $\hbar_\theta$ and $\hbar_\varphi$, the so-called $\hbar$-curve. This is selected in such a way that it controls and converges the series solution. The probable section of $\hbar$ can be found by plotting $\hbar$-curves of $f''(0)$, $\theta'(0)$ and $\varphi'(0)$ for 20th order HAM approximated solution. The valid regions of $\hbar$ are $-1.7 < \hbar_f < 0.1$, $-2.1 < \hbar_\theta < 0.1$ and $-1.5 < \hbar_\varphi < 0.1$, and it is plotted in Figures 2 and 3. The comparison of HAM and numerical methods has been shown graphically in Figures 4–6 and numerically in Tables 1–3. The behavior of the thermophysical parameters involved in non-dimensional velocities, temperature, and concentration field is discussed in Figures 7–21.

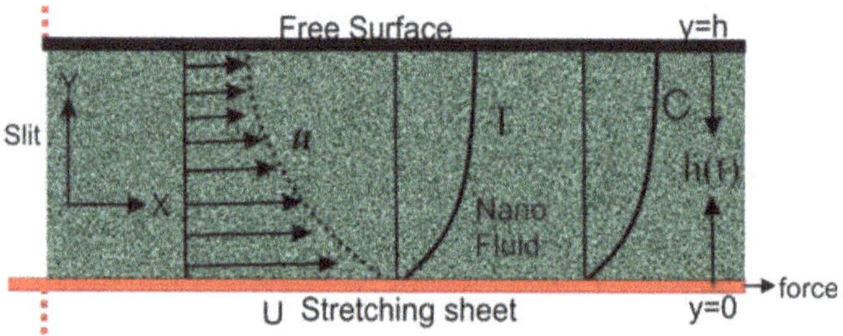

Figure 1. Geometry of the problem

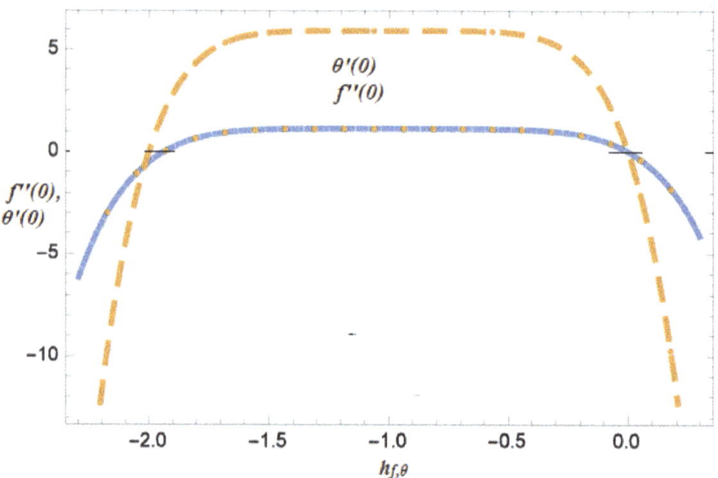

Figure 2. The combined graph of $\hbar$-curves $f''(0)$ $\theta'(0)$, $P_r = 10$, $B_r = 0.8$, $N_r = 0.8$, $D_u = 0.8$, $S_c = 0.4$, $\varepsilon = 0.8$, $S_r = 0.4$, $\lambda = 0.8$, $\Lambda = 1$, $\beta = 1$, $S = 0.3$.

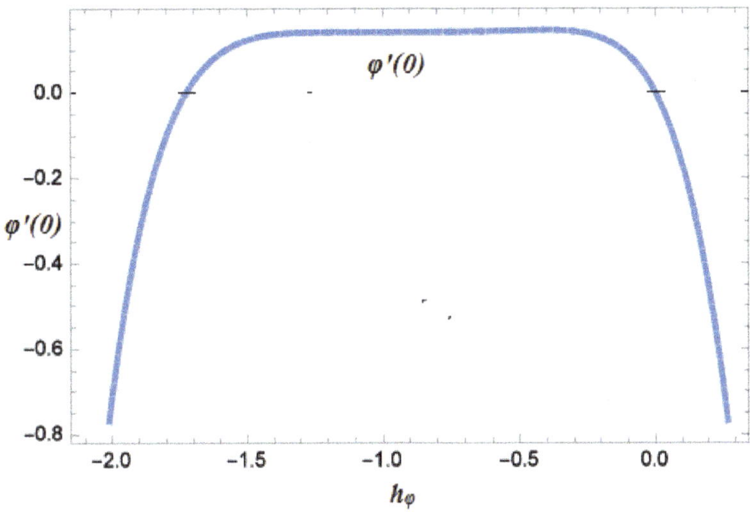

Figure 3. The graph of $\hbar$-curve $\varphi'(0)$, $P_r = 10$, $B_r = 0.8$, $N_r = 0.8$, $D_u = 0.8$, $S_c = 0.4$, $\varepsilon = 0.8$, $S_r = 0.4$, $\lambda = 0.8$, $\Lambda = 1$, $\beta = 1$, $S = 0.3$.

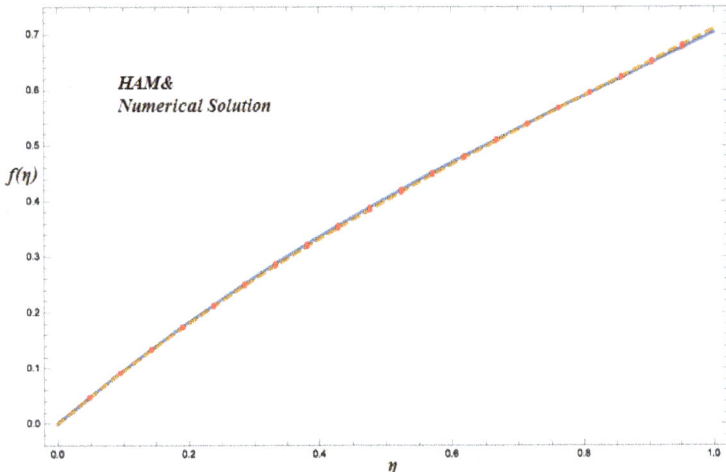

Figure 4. The comparison between HAM and numerical solutions for velocity profile $f(\eta)$, when $\hbar = -0.28$, $P_r = 10$, $B_r = 0.1$, $N_r = 0.1$, $D_u = 0.1$, $S_c = 0.1$, $\varepsilon = 0.1$, $S_r = 0.1$, $\lambda = 0.1$, $\Lambda = 0.1$, $\beta = 1$, $S = 0.1$.

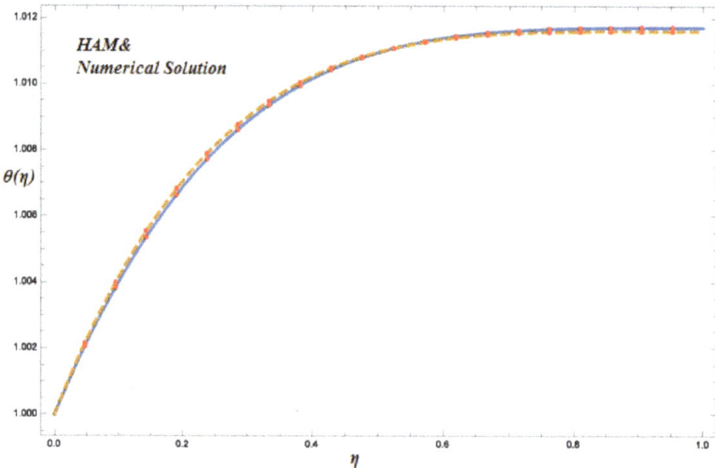

Figure 5. The comparison between HAM and numerical solutions for temperature fields $\theta(\eta)$, when $\hbar = -0.45$, $P_r = 10$, $B_r = 0.7$, $N_r = 0.3$, $D_u = 0.3$, $S_c = 0.9$, $\varepsilon = 0.9$, $S_r = 0.1$, $\lambda = 0.1$, $\Lambda = 0.1$, $\beta = 1$, $S = 0.2$.

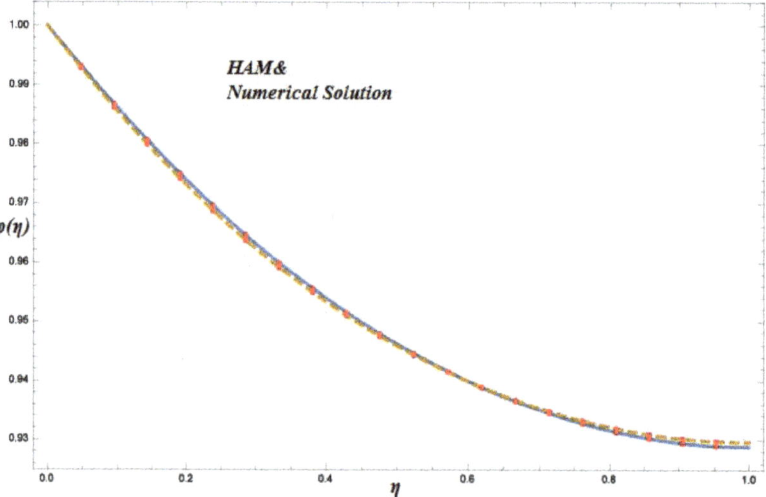

Figure 6. The comparison between HAM and numerical solutions for concentration fields $\varphi(\eta)$, when $\hbar = -0.25$, $P_r = 10$, $B_r = 0.1$, $N_r = 0.1$, $D_u = 0.1$, $S_c = 0.1$, $\varepsilon = 0.1$, $S_r = 0.1$, $\lambda = 0.1$, $\Lambda = 0.1$, $\beta = 1$, $S = 0.1$.

Table 1. Comparison between HAM and numerical solutions for velocity field $f(\eta)$ when $\hbar = -0.28$, $P_r = 10$, $B_r = 0.1$, $N_r = 0.1$, $D_u = 0.1$, $S_c = 0.1$, $\varepsilon = 0.1$, $S_r = 0.1$, $\lambda = 0.1$, $\Lambda = 0.1$, $\beta = 1$, $S = 0.1$.

η	HAM solution Approximation $f(\eta)$	Numerical Solution NN	Absolute Error
0	0.000000	0.000000	0.0
0.1	0.0953944	0.0946811	7.1×10^{-4}
0.2	0.182398	0.180273	2.1×10^{-3}
0.3	0.262182	0.258722	3.4×10^{-3}
0.4	0.335840	0.331586	4.3×10^{-3}
0.5	0.404397	0.400129	4.2×10^{-3}
0.6	0.468820	0.465392	3.4×10^{-3}
0.7	0.530019	0.528247	1.8×10^{-3}
0.8	0.588856	0.589430	5.7×10^{-4}
0.9	0.646152	0.649575	3.4×10^{-3}
1	0.702691	0.709230	6.5×10^{-3}

Table 2. Comparison between HAM and numerical solutions are shown for temperature field $\theta(\eta)$ when $\hbar = -0.45$, $P_r = 10$, $B_r = 0.1$, $N_r = 0.1$, $D_u = 0.1$, $S_c = 0.1$, $\varepsilon = 0.1$, $S_r = 0.1$, $\lambda = 0.1$, $\Lambda = 0.1$, $\beta = 1$, $S = 0.1$.

η	HAM Solution of $\theta(\eta)$	Numerical Solution NN	Absolute Error
0	1.0000	1.00000	0.00000
0.1	1.004	1.00417	1.7×10^{-4}
0.2	1.00688	1.00706	1.9×10^{-4}
0.3	1.00886	1.009	1.8×10^{-3}
0.4	1.01015	1.01023	1.6×10^{-3}
0.5	1.01095	1.01096	7.1×10^{-4}
0.6	1.01139	1.01135	4.2×10^{-4}
0.7	1.0116	1.01153	2.5×10^{-4}
0.8	1.01168	1.01159	1.6×10^{-4}
0.9	1.0117	1.01159	1.1×10^{-4}
1	1.01171	1.01159	1.1×10^{-4}

Table 3. Comparison between HAM and numerical solutions are shown for concentration field $\varphi(\eta)$ when $\hbar = -0.25$, $P_r = 10$, $B_r = 0.1$, $N_r = 0.1$, $D_u = 0.1$, $S_c = 0.1$, $\varepsilon = 0.1$, $S_r = 0.1$, $\lambda = 0.1$, $\Lambda = 0.1$, $\beta = 1$, $S = 0.1$.

η	HAM Solution $\varphi(\eta)$	Numerical Solution NN	Absolute Error
0	1.00000	1.000000	0.000000
0.1	0.986139	0.985513	6.3×10^{-4}
0.2	0.973868	0.973001	8.7×10^{-4}
0.3	0.963145	0.962308	8.4×10^{-4}
0.4	0.953932	0.953301	6.3×10^{-4}
0.5	0.946195	0.945867	3.3×10^{-4}
0.6	0.939906	0.939913	7.5×10^{-6}
0.7	0.935041	0.935364	3.2×10^{-4}
0.8	0.931582	0.93216	5.8×10^{-4}
0.9	0.929513	0.930258	7.4×10^{-4}
1	0.928825	0.929628	8.0×10^{-4}

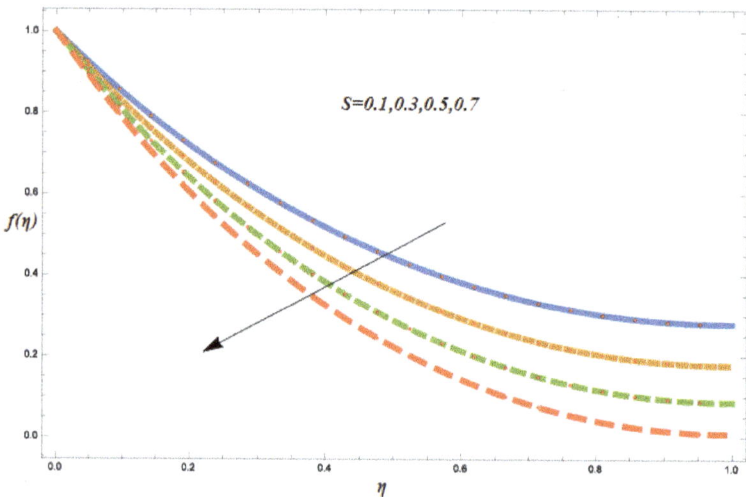

Figure 7. Variants in velocity field $f(\eta)$ for various values of S, when $\hbar = -0.25$, $P_r = 10$, $D_u = 0.7$, $S_c = 0.7$, $\lambda = 0.7$, $\Lambda = 0.7$, $\beta = 1$.

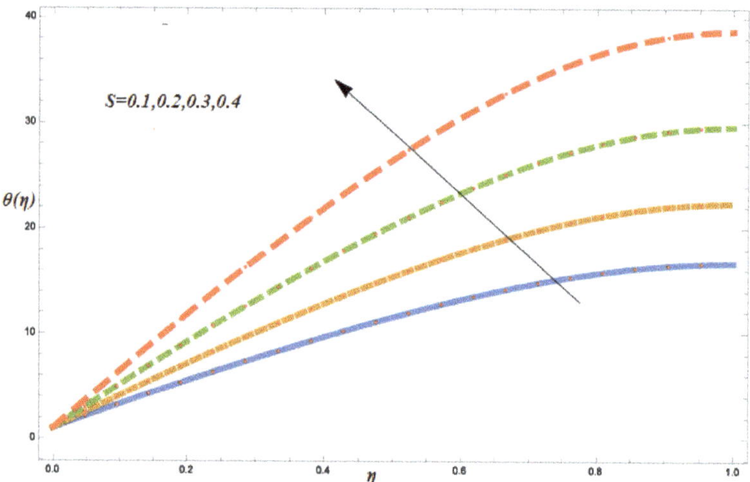

Figure 8. The variation of temperature scale gradient $\theta(\eta)$ for different quantities of S, when $\hbar = -0.25$, $P_r = 10$, $B_r = 0.7$, $N_r = 0.7$, $D_u = 0.7$, $S_c = 0.7$, $\varepsilon = 0.7$, $S_r = 0.7$, $\lambda = 0.7$, $\Lambda = 0.7$, $\beta = 1$.

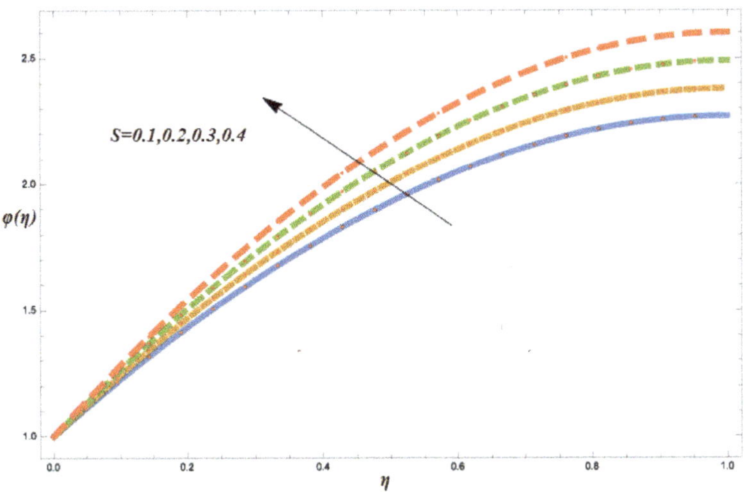

Figure 9. Variations in concentration field $\varphi(\eta)$ occur for different numbers of S, when $\hbar = -0.25$, $P_r = 10$, $B_r = 0.7$, $N_r = 0.7$, $D_u = 0.7$, $S_c = 0.7$, $\varepsilon = 0.7$, $S_r = 0.7$, $\lambda = 0.7$, $\Lambda = 0.7$, $\beta = 1$.

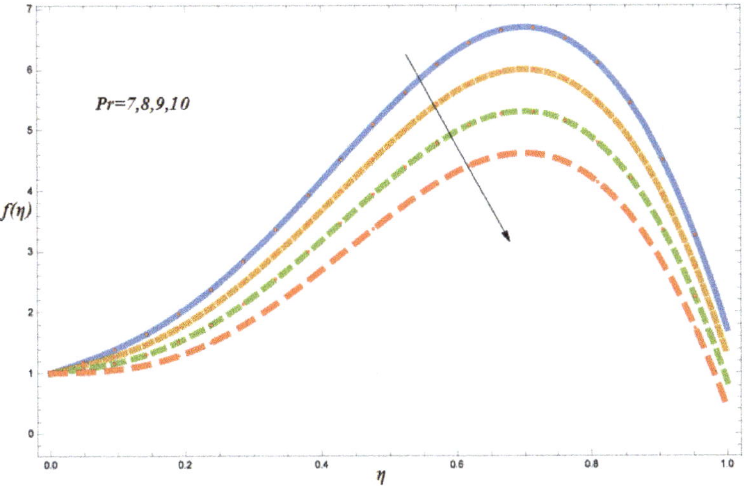

Figure 10. Variation in velocity field $f(\eta)$ for various values of P_r, when $\hbar = -0.25$, $D_u = 0.7$, $S_c = 0.7$, $\lambda = 0.7$, $\Lambda = 0.7$, $\beta = 1$, $S = 0.7$.

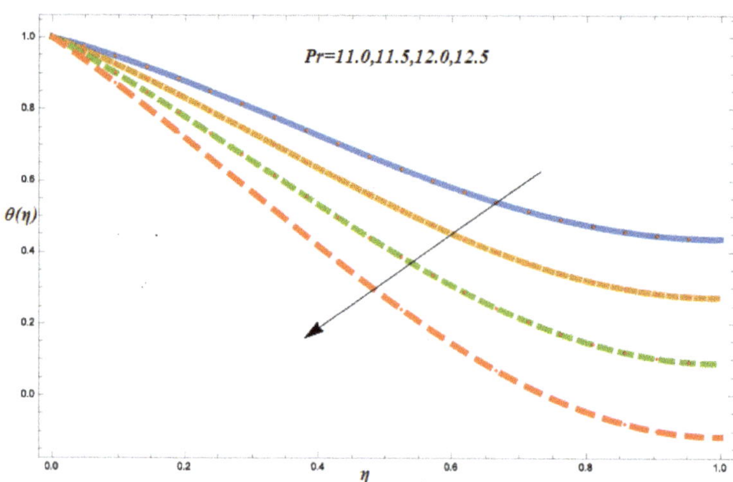

Figure 11. The variation of temperature scale gradient $\theta(\eta)$ for different values of P_r, when $\hbar = -0.25$, $S = 0.7$, $B_r = 0.7$, $N_r = 0.7$, $D_u = 0.7$, $S_c = 0.7$, $\varepsilon = 0.7$, $S_r = 0.7$, $\lambda = 0.7$, $\Lambda = 0.7$, $\beta = 1$.

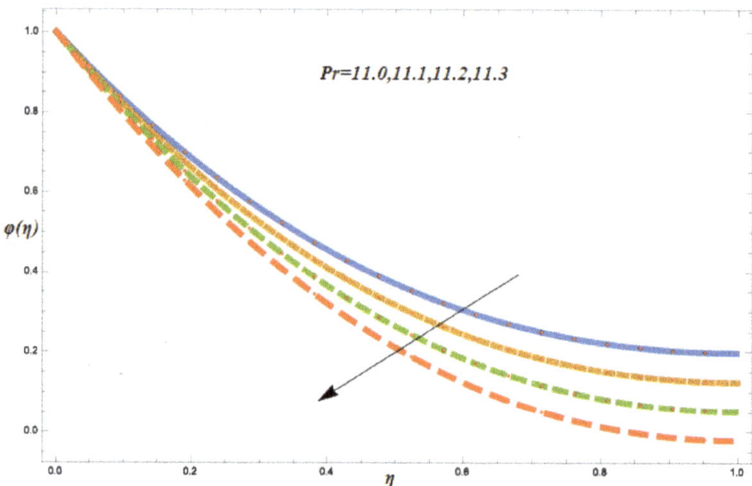

Figure 12. Variations in concentration field $\varphi(\eta)$ occur for different values of Pr, when $\hbar = -0.25$, $S = 0.7$, $B_r = 0.7$, $N_r = 0.7$, $D_u = 0.7$, $S_c = 0.7$, $\varepsilon = 0.7$, $S_r = 0.7$, $\lambda = 0.7$, $\Lambda = 0.7$, $\beta = 1$.

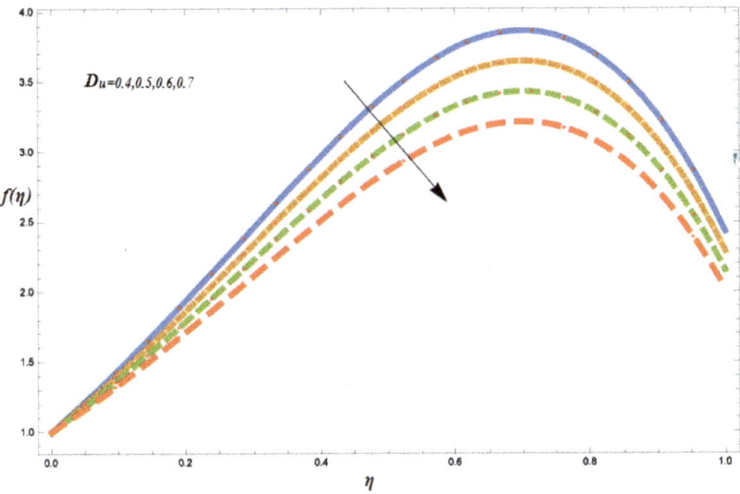

Figure 13. Variations in velocity field $f(\eta)$ for various values of D_u, when $\hbar = -0.25$, $P_r = 10$, $S_c = 0.7$, $\lambda = 0.7$, $\Lambda = 0.7$, $\beta = 1$, $S = 0.7$.

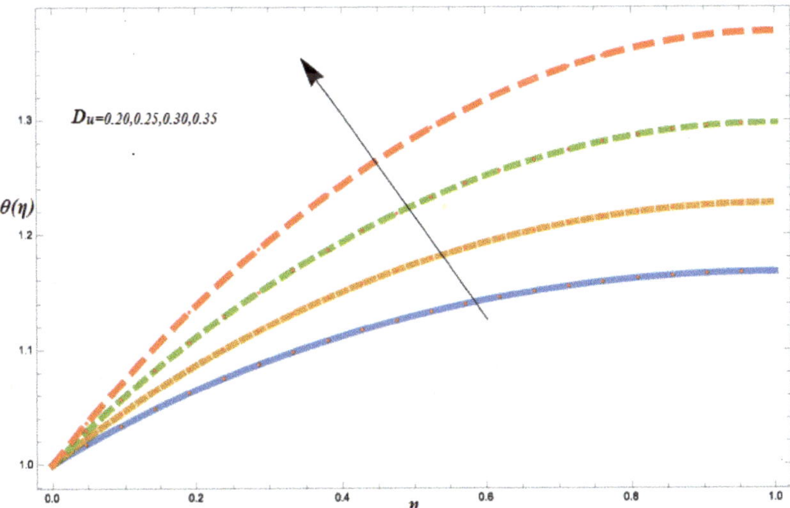

Figure 14. The variation of temperature scale gradient $\theta(\eta)$ for different values of D_u, when $\hbar = -0.25$, $S = 0.7$, $B_r = 0.7$, $N_r = 0.7$, $P_r = 10$, $S_c = 0.7$, $\varepsilon = 0.7$, $S_r = 0.7$, $\lambda = 0.7$, $\Lambda = 0.7$, $\beta = 1$.

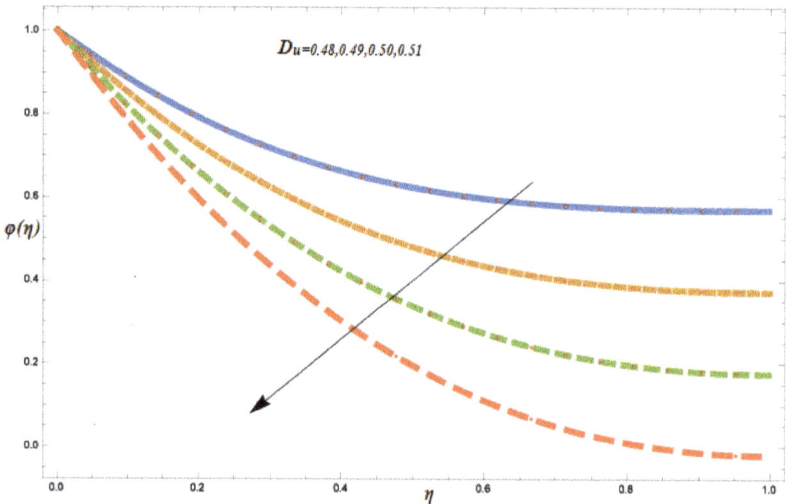

Figure 15. Variations in concentration field φ(η) occur for different values of D_u, when $\hbar = -0.25$, $S = 0.7$, $B_r = 0.7$, $N_r = 0.7$, $P_r = 10$, $S_c = 0.7$, $\varepsilon = 0.7$, $S_r = 0.7$, $\lambda = 0.7$, $\Lambda = 0.7$, $\beta = 1$.

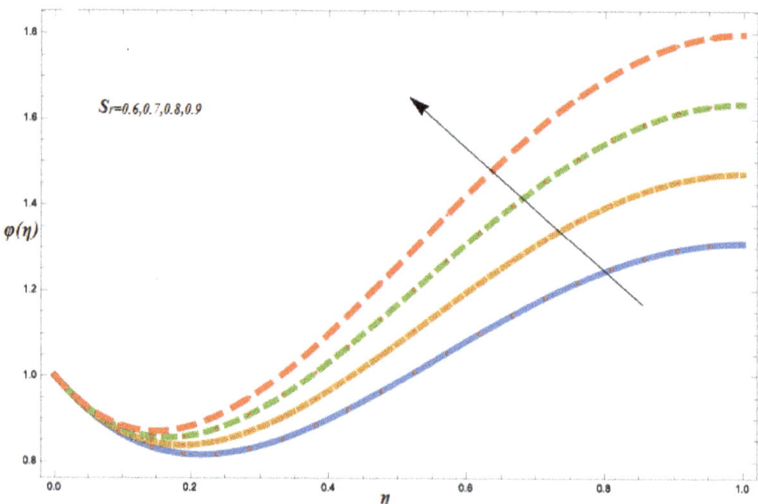

Figure 16. Variations in concentration field φ(η) occur for different values of S_r, when $\hbar = -0.25$, $S = 0.7$, $B_r = 0.7$, $N_r = 0.7$, $P_r = 10$, $S_c = 0.7$, $\varepsilon = 0.7$, $D_u = 0.7$, $\lambda = 0.7$, $\Lambda = 0.7$, $\beta = 1$.

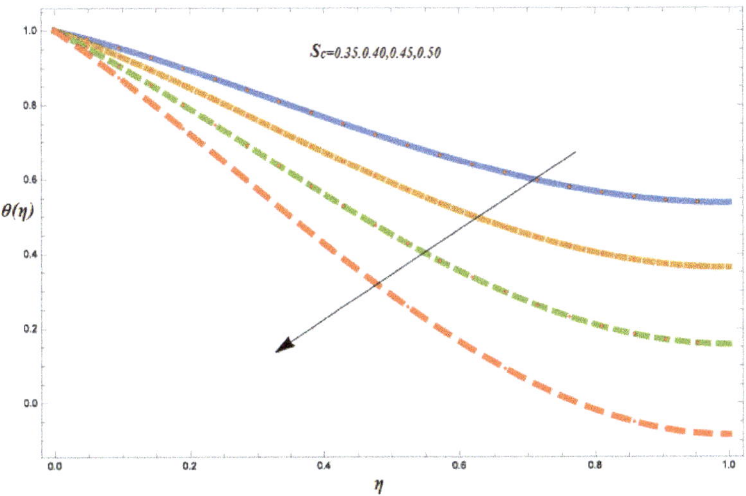

Figure 17. The variation of temperature scale gradient $\theta(\eta)$ for different values of S_c, when $\hbar = -0.25$, $S = 0.7$, $B_r = 0.7$, $N_r = 0.7$, $P_r = 10$, $D_u = 0.7$, $\varepsilon = 0.7$, $S_r = 0.7$, $\lambda = 0.7$, $\Lambda = 0.7$, $\beta = 1$.

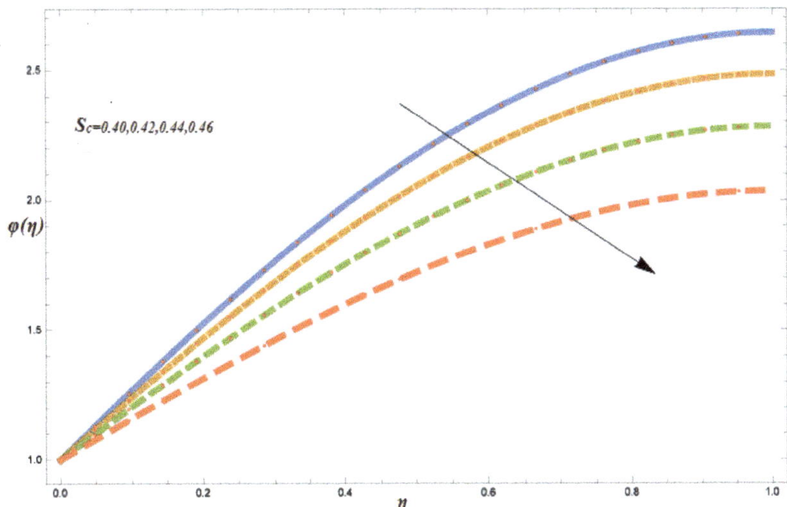

Figure 18. Variations in concentration field $\varphi(\eta)$ occur for different values of S_c, when $\hbar = -0.25$, $S = 0.7$, $B_r = 0.7$, $N_r = 0.7$, $P_r = 10$, $S_r = 0.7$, $\varepsilon = 0.7$, $D_u = 0.7$, $\lambda = 0.7$, $\Lambda = 0.7$, $\beta = 1$.

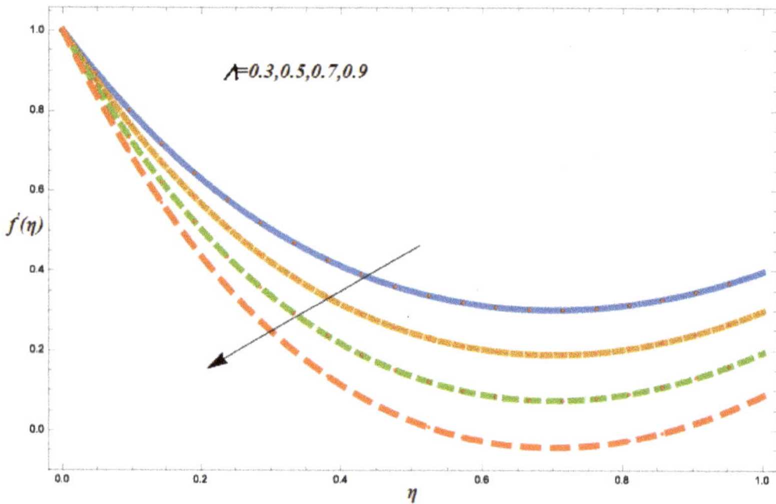

Figure 19. Variations in velocity field $f'(\eta)$ for various values of Λ, when $\hbar = -0.25$, $P_r = 10$, $S_c = 0.7$, $\lambda = 0.7$, $D_u = 0.7$, $\beta = 1$, $S = 0.7$.

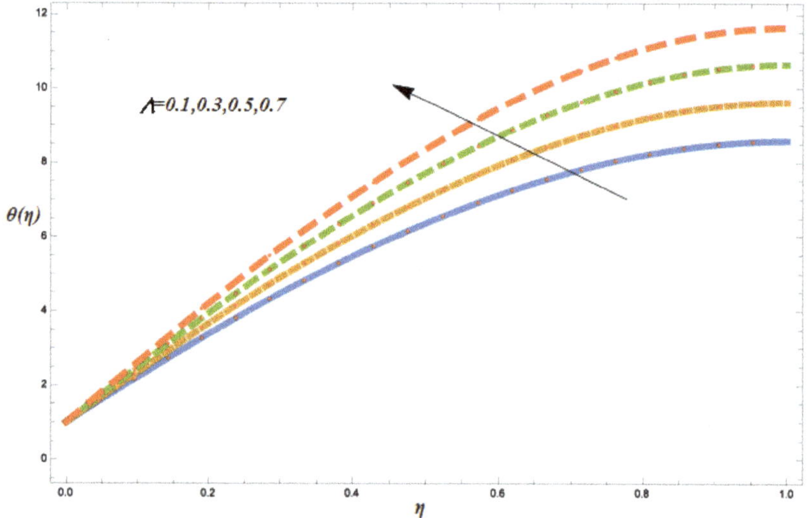

Figure 20. The variation of temperature scale gradient $\theta(\eta)$ for different values of Λ, when $\hbar = -0.25$, $S = 0.7$, $B_r = 0.7$, $N_r = 0.7$, $P_r = 10$, $D_u = 0.7$, $\varepsilon = 0.7$, $S_r = 0.7$, $\lambda = 0.7$, $S_c = 0.7$, $\beta = 1$.

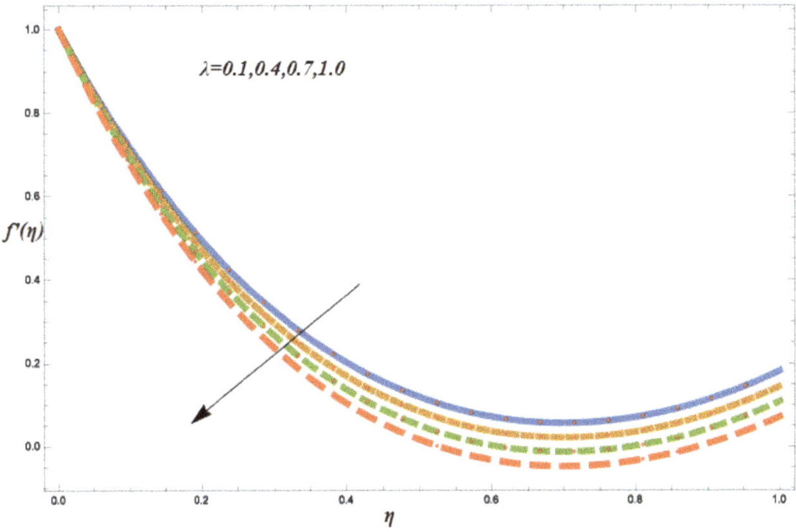

Figure 21. Variations in velocity field $f'(\eta)$ for various values of λ, when $\hbar = -0.25$, $P_r = 10$, $S_c = 0.7$, $\Lambda = 0.7$, $D_u = 0.7$, $\beta = 1$, $S = 0.7$.

4. Discussion

In this work, numerical values are assigned to the physical parameters involved in the velocity, temperature, and concentration profiles. The numerical outcomes for velocity, temperature, and concentration profiles are presented in this section. An efficient numerical method called the ND-solve method has been used to solve the transformed Equations (8)–(10) subject to the boundary conditions in Equation (11). The paper examined the effects of governing parameters on the transient velocity profile, temperature profile, and concentration profile. For this purpose the SRM approach has been applied for various values of flow controlling parameters $S = 0.7$, $P_r = 10$, $D_u = 0.7$, $S_r = 0.7$, $S_c = 0.7$, $\Lambda = 0.7$, $\lambda = 0.7$ to obtain a clear insight into the physics of the problem. Therefore, all the graphs and tables correspond to the values above and the rest will be mentioned. The behavior of the non-dimensional unsteady parameter S for velocities, temperature, and concentration field during fluid motion is studied in Figures 7–9. The unsteady parameter S is inversely related to the stretching constant of the velocity field, whereas it is directly related to the stretching constants of the temperature and concentration fields. Therefore, by increasing the values of S the value of the velocity field is decreased while the values of the temperature and concentration fields increase. An increase in P_r leads to an increase in kinematic viscosity and a decrease in velocity. The reason is that the rise in viscosity tends to increase the resistance force and as a result the velocity profile descends (Figure 10). Figure 11 shows the effect of Prandtl number P_r in temperature fields; the same effect is observed for velocity fields. The thermal diffusion falls with the rise in Prandtl number P_r and as a result the thermal boundary layer becomes thinner and the temperature decreases. This variation in thermal diffusivity is due to the difference of temperature fields; the fluid is highly conductive. Therefore, a fluid with greater P_r and larger heat capacity increases the heat transfer, the same as in [21]. This variation in thermal diffusivity is due to the difference of temperature fields. The same effect for concentration field is exposed in Figure 12. The behavior of Dufour number D_u is discussed in Figures 13–15. The Dufour number is actually the ratio of temperature and concentration difference. The Soret effect is a mass flux due to a temperature gradient and the Dufour effect is enthalpy flux due to a concentration gradient and appears in the energy equation. It was also observed that the effect of D_u and S_r on the temperature and concentration fields is opposite. In Figure 13 it is shown that increasing the value of Dufour number D_u decreases the velocity profile. Since the Dufour number D_u has an inverse relationship with thermal diffusion, we conclude that the falls in fluid velocity are due to the smaller thermal diffusion. However, it is clear in Figure 14 that the temperature field increases for greater values of D_u. Physically, the Dufour effect has a direct relationship with the concentration gradient of energy flux and, as a result, temperature increases for larger values of D_u. The concentration field decreases with increasing values of the Dufour number D_u, as shown in Figure 15. The Soret number is the reciprocal ratio of the Dufour number; due to this property the reverse physical behavior of the Soret S_r and Dufour numbers D_u has been noticed in the concentration field and is shown in Figure 16. The effects of Schmidt number S_c on the temperature field and concentration field are discussed in Figures 17 and 18, respectively. Figure 17 exhibits the effect of Schmidt number on temperature fields: an increase in the value of S_c increases the temperature field. The influence of the Schmidt number S_c on the concentration field is shown in Figure 18. Increasing the Schmidt number S_c reduces the concentration boundary layer, because the increase in Schmidt number S_c means lower molecular diffusivity, which decreases the concentration boundary layer. It is observed that an increase in S_c leads to a decrease in the heat transfer rate at the surface. The variable viscosity parameter Λ plays a significant role in the flow, as shown in Figures 19 and 20. The viscosity of the fluid is directly related to the cohesive and adhesive forces. So by increasing the cohesive and adhesive forces, the fluid resistance is increased, which results in a decrease in the fluid velocity $f'(\eta)$, as shown in Figure 19. On the other hand, it is inversely related to the temperature field, as shown in Figure 20, i.e., increasing the temperature of the fluid decreases the viscosity. This is because increasing the values of temperature causes the cohesive and adhesive forces of the fluid to become weaker. Due to this, the thickness of the fluid decreases. The effect of the Williamson parameter λ on the velocity profile is exhibited in Figure 21. The velocity reduces when λ is augmented because a rise in relaxation time causes higher resistance in the fluid flow and as a result reduces the velocity field. The comparison of HAM and numerical solutions for the velocity, temperature, and concentration fields are shown in Tables 1–3 and a closed agreement between these two methods has been observed.

5. Conclusions

The governing equations are modeled and solved for the thin film flow of nanofluid. A non-Newtonian Williamson fluid is used as a base fluid in the presence of thermal radiation. The nonlinear coupled equations have been solved using HAM and are compared with the numerical solutions.

The key points of this work are:

- The variable effects of the fluid properties on the flow of a Williamson nanofluid are plotted through graphs and tables.
- The Dufour and Soret effects during thin film nanofluid motion are considered in the presence of thermal radiation.
- Experimental values of the Prandtl number have been used to produce the most accurate results for the Williamson nanofluid.
- The accuracy of the HAM results has been verified via numerical solutions.

Author Contributions: Taza Gul and Waris Khan modeled the problem and solved it; Muhammad Idrees, Waris Khan and L.C.C. Dennis contributed to the discussion of the problem; Saeed Islam, Ilyas Khan, L.C.C. Dennis contributed in the English corrections, All the authors read and approved the final manuscript.

Conflicts of Interest: The authors declare no conflict of interest.

Nomenclature

x, y	Cartesian coordinates
U_0	Stretching velocity
b	Stretching velocity constraint
T_s	Temperature field

C_s	Concentration filed
T_0	Surface temperature
T_{ref}	Reference temperature
C_{ref}	Reference concentration
$\mu(T)$	Variable viscosity
μ_0	Fluid viscosity at reference temperature
γ	Dependency strength
$K(T)$	Temperature-dependent thermal conductivity
ε	Variable thermal conductivity parameter
υ	Kinematics viscosity
Γ	Time parameter
u, v	Velocity components
T	Temperature field
C	Concentration field
ρ	Fluid density
C_p	Specific heat
$h(t)$	Liquid film thickness
q_r	Radiative heat fluctuation
σ	Stefan–Boltzmann constant
D_m	Concentration molecular diffusivity
T_m	Mean temperature
k_T	Thermal diffusion ratio
k	Thermal conductivity of the liquid film
ψ	Stream function
β	Non-dimensional thickness of the nano liquid film
Λ	Variable viscosity parameter
Pr	Prandtl number
S	Non-dimensional measure of unsteadiness
D_u	Dufour number
S_c	Schmidt number
S_r	Soret number
R	Radiation constant
Br	Brinkman number
N_r	Thermal radiation parameter
λ	Williamson fluid constant
C_s	Concentration vulnerability

References

1. Chakrabarti, A.; Gupta, A.S. Hydromagnetic flow and heat transfer over a stretching sheet. *Quart. J. Appl. Math.* **1979**, *37*, 73–78.
2. Sakiadis, B.C. Boundary layer behaviour on continuous moving solid surfaces. I. Boundary layer equations for two-dimensional and axisymmetric flow. II. Boundary layer on a continuous flat surface. III. Boundary layer on a continuous cylindrical surface. *Am. Inst. Chem. Eng. J.* **1961**, *7*, 26–28. [CrossRef]
3. Crane, L.J. Flow past a stretching sheet. *Z. Appl. Math. Phys* **1970**, *21*, 645–647. [CrossRef]
4. Gupta, P.S.; Gupta, A.S. Heat and mass transfer on a stretching sheet with suction or blowing. *Can. J. Chem. Eng.* **1977**, *55*, 744–746. [CrossRef]
5. Elbashbeshy, E.M.A. Heat transfer over a stretching surface with variable surface a heat flux. *J. Phys. D* **1998**, *31*, 1951–1954. [CrossRef]
6. Abd El-Aziz, M. Radiation effect on the flow and heat transfer over an unsteady stretching sheet. *Int. Commun. Heat Mass Transf.* **2009**, *36*, 521–524. [CrossRef]
7. Mukhopadyay, S. Effect of thermal radiation on unsteady mixed convection flow and heat treansfer over a porous stretching surface in porous medium. *Int. Commun. Heat Mass Transf.* **2009**, *52*, 3261–3265. [CrossRef]

8. Shateyi, S.; Motsa, S.S. Thermal radiation effects on heat and mass transfer over an unsteady stretching surface. *Math. Probl. Eng.* **2009**, *2009*, 13. [CrossRef]
9. Abd El-Aziz, M. Thermal-diffusion and diffusion-thermo effects on combined heat and mass transfer by hydromagnetic three-dimensional free convection over a permeable stretching surface with radiation. *Phys. Lett.* **2007**, *372*, 263–272. [CrossRef]
10. Hady, F.M.; Ibrahim, F.S.; Abdel-Gaied, S.M.; Eid, M.R. Radiation effect on viscous flow of a nanofluid and heat transfer over a nonlinearly stretching sheet. *Nanoscale Res. Lett.* **2012**, *7*, 229. [CrossRef] [PubMed]
11. Pavlov, K.B. Magnetohydromagnetic flow of an incompressible viscous fluid caused by deformation of a surface. *Magn. Gidrodin.* **1974**, *4*, 146–148.
12. Bianco, V.; Manca, O.; Nardini, S. Second Law Analysis of Al2O3-Water Nanofluid Turbulent Forced Convection in a Circular Cross Section Tube with Constant Wall Temperature. *Adv. Mech. Eng.* **2013**, 920278. [CrossRef]
13. Nadeem, S.; Haq, R.U.; Noreen, S.A.; Khan, Z.H. MHD three-dimensional Casson fluid flow past a porous linearly stretching sheet. *Alex. Eng. J.* **2013**, *52*, 577–582. [CrossRef]
14. Nadeem, S.; Ul Haq, R.; Lee, C. MHD flow of a Casson fluid over an exponentially shrinking sheet. *Sci. Iran.* **2012**, *19*, 1550–1553. [CrossRef]
15. Nadeem, S.; Ul Haq, R.; Akbar, N.S.; Lee, C.; Khan, Z.H. Numerical Study of Boundary Layer Flow and Heat Transfer of Oldroyd-B Nanofluid towards a Stretching Sheet. *PLoS ONE* **2013**, *8*, e69811. [CrossRef] [PubMed]
16. Nadeem, S.; Ul Haq, R.; Khan, Z.H. Numerical study of MHD boundary layer flow of a Maxwell fluid past a stretching sheet in the presence of nanoparticles. *J. Taiwan Inst. Chem. Eng.* **2014**, *45*, 121–126. [CrossRef]
17. Elbashbeshy, E.M.A.; Bazid, M.A.A. Heat transfer over an unsteady stretching surface with internal heat generation. *Appl. Math. Comput.* **2003**, *138*, 239–245. [CrossRef]
18. Grubka, L.J.; Bobba, K.M. Heat transfer characteristics of a continuous stretching surface with variable temperature. *J. Heat Transf.* **1985**, *107*, 248–250. [CrossRef]
19. Chen, C.K.; Char, M.I. Heat transfer of a continuous, stretching surface with suction or blowing. *J. Math. Anal. Appl.* **1988**, *135*, 568–580. [CrossRef]
20. Pop, I.; Gorla, R.S.R.; Rashidi, M. The effect of variable viscosity on flow and heat transfer to a continuous moving flat plate. *Int. J. Eng. Sci.* **1992**, *30*, 1–6. [CrossRef]
21. Pantokratoras, A. Further results on the variable viscosity on flow and heat transfer to a continuous moving flat plate. *Int. J. Eng. Sci.* **2004**, *42*, 1891–1896. [CrossRef]
22. Abel, M.S.; Khan, S.K.; Prasad, K.V. Study of visco-elastic fluid flow and heat transfer over a stretching sheet with variable viscosity. *Int. J. Non-Linear Mech.* **2002**, *37*, 81–88. [CrossRef]
23. Makinde, O.D.; Mishra, S.R. On Stagnation Point Flow of Variable Viscosity Nano fluids Past a Stretching Surface with Radiative Heat. *Int. J. Appl. Comput. Math* **2015**. [CrossRef]
24. Mukhopadhyay, S.; Layek, G.C.; Samad, S.K.A. Study of MHD boundary layer flow over a heated stretching sheet with variable viscosity. *Int. J. Heat Mass Transf.* **2005**, *48*, 4460–4466. [CrossRef]
25. Hayat, T.; Muhammad, T.; Shehzad, S.A.; Alsaedi, A. Soret and Dufour effects in three-dimensional flow over an exponentially stretching surface with porous medium, chemical reaction and heat source/sink. *Int. J. Numer. Methods Heat Fluid Flow* **2015**, *25*, 762–781. [CrossRef]
26. Alam, M.S.; Ferdows, M.; Ota, M.; Maleque, M.A. Dufour and Soret effects on steady free convection and mass transfer flow past a semi-infinite vertical porous plate in a porous medium. *Int. J. Appl. Mech. Eng.* **2006**, *11*, 535–545.
27. Kafoussias, N.G.; Williams, E.W. Thermal-diffusion and diffusion thermo effects on mixed free-forced convective and mass transfer boundary layer flow with temperature dependent viscosity. *Int. J. Eng. Sci.* **1995**, *33*, 1369–1384. [CrossRef]
28. Chamkha, A.J.; Ben-Nakhi, A. MHD mixed convection-radiation interaction along a permeable surface immersed in a porous medium in the presence of Soret and Dufour's effects. *Heat Mass Transf.* **2008**, *44*, 845–856. [CrossRef]
29. Afify, A.A. Similarity solution in MHD Effects of thermal diffusion and diffusion thermo on free convective heat and mass transfer over a stretching surface considering suction or injection. *Commun. Nonlinear Sci. Numer. Simul.* **2009**, *14*, 2202–2214. [CrossRef]

30. Be'g, O.A.; Bakier, A.Y.; Prasad, V.R. Numerical study of free convection magnetohydrodynamic heat and mass transfer from a stretching surface to a saturated porous medium with Soret and Dufour effects. *Comput. Mater. Sci.* **2009**, *46*, 57–65. [CrossRef]
31. El-Kabeir, S.M.M.; Chamkha, A.J.; Rashad, A.M.; Al-Mudhaf, H.F. Soret and Dufour effects on heat and mass transfer by non-Darcy natural convection from a permeable sphere embedded in a high porosity medium with chemically-reactive species. *Int. J. Energy Technol.* **2010**, *2*, 1–10.
32. Pal, D.; Mondal, H. Effects of Soret Dufour, chemical reaction and thermal radiation on MHD non-Darcy unsteady mixed convective heat and mass transfer over a stretching sheet. *Commun. Nonlinear Sci. Numer. Simul.* **2011**, *16*, 1942–1958. [CrossRef]
33. Khan, Y.; Wu, Q.; Faraz, N.; Yildirim, A. The effects of variable viscosity and thermal conductivity on a thin film flow over a shrinking/stretching sheet. *Comput. Math. Appl.* **2011**, *61*, 3391–3399.
34. Aziz, R.C.; Hashim, I.; Alomari, A.K. Thin film flow and heat transfer on an unsteady stretching sheet with internal heating. *Meccanica* **2011**, *46*, 349–357. [CrossRef]
35. Qasim, M.; Khan, Z.H.; Lopez, R.J.; Khan, W.A. Heat and mass transfer in nanofluid over an unsteady stretching sheet using Buongiorno's model. *Eur. Phys. J. Plus* **2016**, *131*, 1–16. [CrossRef]
36. Prashan, G.M.; Jagdish, T.; Abel, M.S. Thin film flow and heat transfer on an unsteady stretching sheet with thermal radiation, internal heating in presence of external magnetic field. *Phys. Flu. Dyn.* **2016**, *3*, 1–16.
37. Ellahi, R.; Hassan, M.; Zeeshan, A. Aggregation effects on water base Al_2O_3—Nanofluid over permeable wedge in mixed convection. *Asia-Pac. J. Chem. Eng.* **2016**, *11*, 179–186. [CrossRef]
38. Akbar, N.S.; Raza, M.; Ellahi, R. CNT suspended CuO + H_2O nano fluid and energy analysis for the peristaltic flow in a permeable channel. *Alex. Eng. J.* **2015**, *54*, 623–633. [CrossRef]
39. Akbar, N.S.; Raza, M.; Ellahi, R. Copper oxide nanoparticles analysis with water as base fluid for peristaltic flow in permeable tube with heat transfer. *Comput. Methods Progr. Biomed.* **2016**, *130*, 22–30. [CrossRef] [PubMed]
40. Shehzad, N.; Zeeshan, A.; Ellahi, R.; Vafai, K. Convective heat transfer of nanofluid in a wavy channel: Buongiorno's mathematical model. *J. Mol. Liq.* **2016**, *222*, 446–455. [CrossRef]
41. Zeeshan, A.; Hassan, M.; Ellahi, R.; Nawaz, M. Shape effect of nanosize particles in unsteady mixed convection flow of nanofluid over disk with entropy generation. *J. Process Mech. Eng.* **2016**, 1–9. [CrossRef]
42. Liao, S.J. *Homotopy Analysis Method in Nonlinear Differential Equations*; Higher education press: Beijing, China, 2012.
43. Liao, S. *Beyond Perturbation: Introduction to the Homotopy Analysis Method*; Chapman & Hall/CRC: Boca Raton, FL, USA, 2003.
44. Liao, S.J. An optimal homotopy-analysis approach for strongly nonlinear differential equations. *Commun. Nonlinear Sci. Numer. Simul.* **2010**, *15*, 2003–2016. [CrossRef]
45. Liao, S. On the homotopy analysis method for nonlinear problems. *Appl. Math. Comput.* **2004**, *147*, 499–513. [CrossRef]
46. Abbasbandy, S.; Shirzadi, A. A new application of the homotopy analysis method: Solving the Sturm—Liouville problems. *Commun. Nonlinear Sci. Numer. Simul.* **2011**, *16*, 112–126. [CrossRef]
47. Abbasbandy, S. Homotopy analysis method for heat radiation equations. *Int. Commun. Heat Mass Transf.* **2007**, *34*, 380–388. [CrossRef]
48. Abbasbandy, S. The application of homotopy analysis method to solve a generalized Hirota-Satsuma coupled KdV equation. *Phys. Lett. A* **2007**, *361*, 478–483. [CrossRef]
49. Das, K. Effects of thermophoresis and thermal radiation on MHD mixed convective heat and mass transfer flow. *Afr. Math. Union Springer-Verl.* **2012**, *24*, 511–524. [CrossRef]
50. Qasim, M. Soret and Dufour effects on the flow of an Erying-Powell fluid over a flat plate with convective boundary condition. *Eur. Phys. J Plus* **2014**, *129*, 1–7. [CrossRef]
51. Mahesh, K.; Gireesha, B.J.; Rama, S.R.G. Heat and Mass Transfer in a Nanofluid Film on an Unsteady Stretching Surface. *J. Nanofluids* **2015**, *4*, 1–8.

Article

Slip Flow and Heat Transfer of Nanofluids over a Porous Plate Embedded in a Porous Medium with Temperature Dependent Viscosity and Thermal Conductivity

Sajid Hussain [1], Asim Aziz [2,*], Taha Aziz [3] and Chaudry Masood Khalique [3]

[1] Department of Mathematics, Capital University of Science and Technology, Islamabad 44000, Pakistan; prsajid@yahoo.com

[2] College of Electrical and Mechanical Engineering, National University of Sciences and Technology, Rawalpindi 46070, Pakistan

[3] International Institute for Symmetry Analysis and Mathematical Modeling, Department of Mathematical Sciences, North-West University, Mafikeng Campus, Private Bag X 2046, Mmabatho 2735, South Africa; tahaaziz77@yahoo.com (T.A.); Masood.Khalique@nwu.ac.za (C.M.K.)

* Correspondence: aaziz@ceme.nust.edu.pk; Tel.: +92-3325-464-647

Academic Editor: Rahmat Ellahi
Received: 20 September 2016; Accepted: 3 November 2016; Published: 14 December 2016

Abstract: It is well known that the best way of convective heat transfer is the flow of nanofluids through a porous medium. In this regard, a mathematical model is presented to study the effects of variable viscosity, thermal conductivity and slip conditions on the steady flow and heat transfer of nanofluids over a porous plate embedded in a porous medium. The nanofluid viscosity and thermal conductivity are assumed to be linear functions of temperature, and the wall slip conditions are employed in terms of shear stress. The similarity transformation technique is used to reduce the governing system of partial differential equations to a system of nonlinear ordinary differential equations (ODEs). The resulting system of ODEs is then solved numerically using the shooting technique. The numerical values obtained for the velocity and temperature profiles, skin friction coefficient and Nusselt's number are presented and discussed through graphs and tables. It is shown that the increase in the permeability of the porous medium, the viscosity of the nanofluid and the velocity slip parameter decrease the momentum and thermal boundary layer thickness and eventually increase the rate of heat transfer.

Keywords: nanofluids; variable viscosity; variable thermal conductivity; partial slip; heat transfer; porous plate

1. Introduction

The heat transfer due to fluid flow is an important factor in problems in industries, such as heat exchangers, the recovery of petroleum resources, fault zones, catalytic reactors, cooling systems, electronic equipment manufacturing, etc. The heat transfer characteristics in the boundary layer are influenced by a number of factors, including flow geometry, the viscosity of a fluid, thermal conductivity, bounding surface characteristics, boundary conditions, flow medium and the orientation and intensity of the applied magnetic field [1–3]. Maxwell first proposed that the thermal conductivity of the fluid can be increased by including solid particles in the flow domain [4]. Following Maxwell, extensive research has been conducted to study the heat transfer characteristics of fluid flow in a porous medium. It is beyond the scope of this work to revisit the vast amount of literature on different Newtonian and non-Newtonian fluids' flow within a porous medium. A comprehensive literature on forced/natural convective heat transfer in porous medium can be found in [5,6].

The introduction of nanofluids by Choi [7] offered new possibilities of heat transfer enhancement, and a number of studies were conducted to study the effects of the thermal properties (mainly thermal conductivity), viscosity and convective heat transfer performance of nanofluids. Experiments performed by Wang et al. [8] and Keblinski et al. [9] showed that the effective thermal conductivity of nanofluids increases under macroscopically stationary conditions. Buongiorno [10] performed a detailed analysis on convective transport in nanofluids. A comprehensive literature survey on transport and heat transfer characteristics of nanofluids was presented in the review articles of Keblinski et al. [11] and Wang and Mujumdar [12]. It has been demonstrated that nanofluids can have significantly better heat transfer characteristics than the conventional fluids depending upon the type, size and concentration of nanoparticles and the nanofluids' transport through the porous media.

Nield and Kuznetsov [13,14] first studied the effects of porous media, thermophoresis and Brownian motion on the convective heat transfer of nanofluids. Sun and Pop [15] found the numerical solution of the steady-state free convection heat transfer behavior of nanofluids inside a triangular enclosure saturated by a porous media. It was observed that the heat transfer rate increases with the increase in nanoparticle volume concentration at a low Rayleigh number, whereas the opposite trend was observed for a high Rayleigh number. Khan and Aziz [16] studied the double-diffusive free convection from a vertical plate to a porous medium saturated with a binary base nanofluid. The influence of the internal heat source on the onset of Darcy–Brinkman convection in a porous layer filled with a nanofluid was presented by Yadav et al. [17]. They showed that the porous medium has stabilizing effects on the modeled system. Khan et al. [18] studied the free convection of nanofluids along a vertical plate in porous media. Servati et al. [19] studied numerically the force convective MHDflow of a nanofluid in a channel partially filled with porous media. The steady mixed convection boundary layer flow of nanofluids past a vertical flat plate embedded in porous media was discussed by Ahmad and Pop [20]. Recently, Cimpean and Pop [21] presented a detailed study on the flow of three different nanofluids (Cu−water,Al_2O_3−water and TiO_2−water) in an inclined channel saturated by a porous media. A review article detailing the literature on the convective heat transfer of nanofluids in porous media and some recent investigations on nanofluids models and related topics can be found in [22–26].

It can be seen from the available literature that limited or no attention has been given to the slip wall condition and the effects of variable thermophysical properties on the flow and heat transfer characteristics of nanofluids. Wall slip has far-reaching implications for many branches of science, engineering and industry. These include rheometric measurements, material processing and fluid transportation [27,28]. Moreover, many processes in engineering occur at high temperature, and it is well known that the thermophysical properties of fluids may change with temperature and become important for the design of reliable equipment, nuclear plants, gas turbines and various propulsion devices or aircraft, missiles, satellites and space vehicles. On the basis of these applications, Khan et al. [29] studied the flow and heat transfer of carbon nanotubes (CNTs) subjected to Navier slip and uniform heat flux boundary conditions. Zheng et al. [30] extended the idea and studied the effects of velocity slip and temperature jump on MHD flow and heat transfer of nanofluids over a porous shrinking sheet. Moreover, Zhenga et al. [31] presented an investigation for the flow and radiation heat transfer of a nanofluid over a porous sheet with velocity slip and temperature jump in a porous medium. Uddin et al. [32] analyzed numerically the g-Jittermixed convective unsteady slip flow of nanofluids past a permeable linear porous sheet embedded in a Darcian porous media with variable viscosity. Noghrehabadi et al. [33] observed the effects of partial slip boundary conditions on the flow and heat transfer of nanofluids. Bhaskar et al. [34] carried out an analysis to investigate the influence of variable thermal conductivity and partial velocity slip on the hydromagnetic two-dimensional boundary layer flow of nanofluids over a porous sheet with a convective boundary condition. Noghrehabadi et al. [35] carried out a study on the effects of variable thermal conductivity and viscosity on the natural convective heat transfer of nanofluids over a vertical plate. Comprehensive studies and lists of important references on the wall slip condition and variable thermophysical properties of nanofluids are presented in [36–41].

In the present work, a mathematical model is presented to study the effects of partial slip, variable viscosity and variable thermal conductivity on steady boundary layer flow of a nanofluid over a porous sheet in a Darcy-type porous medium. The wall slip conditions are employed in terms of shear stress, with viscosity and thermal conductivity as linear functions of temperature. Similarity solutions are obtained, and the reduced system of ordinary differential equations is solved numerically using the shooting method. The numerical results obtained for the velocity and temperature profiles are influenced appreciably by the presence of variable viscosity, variable thermal conductivity, porous medium, velocity and temperature slip and suction/injection parameters. The effects of various parameters on velocity and temperature profiles, as well as skin friction and the rate of heat transfer are presented and discussed through graphs and tables.

2. Mathematical Model of the Problem

We consider the steady two-dimensional laminar boundary layer flow with heat transfer of an incompressible nanofluid over a semi-infinite porous plate in a porous medium. The surface of the plate is at constant temperature T_w and admits the partial slip condition. The viscosity and the thermal conductivity of the nanofluid are considered to vary linearly with temperature. The x-axis is along the surface of the plate, and the y-axis is perpendicular to it. All body forces are neglected, and there is a constant suction/injection velocity V_w at the surface of the plate. The flow far away from the plate is uniform and in the direction parallel to the plate. The velocity and temperature outside the boundary layer are u_∞ and T_∞, respectively. The geometry of the flow model is given in Figure 1.

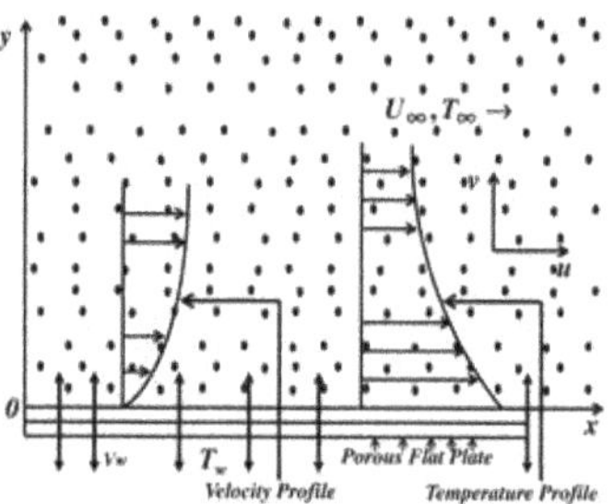

Figure 1. Schematic representation of the geometry.

In view of the above assumptions, the continuity, momentum and energy equations for the flow along with heat transfer are:

$$\frac{\partial u}{\partial x} + \frac{\partial v}{\partial y} = 0, \tag{1}$$

$$u\frac{\partial u}{\partial x} + v\frac{\partial u}{\partial y} = \frac{1}{\rho_{n_f}}\frac{\partial}{\partial y}\left[\mu_{n_f}(T)\frac{\partial u}{\partial y}\right] - \frac{\mu_{n_f}(T)}{\rho_{n_f}k}(u - u_\infty), \tag{2}$$

$$u\frac{\partial T}{\partial x} + v\frac{\partial T}{\partial y} = \frac{1}{(\rho C_p)_{n_f}}\frac{\partial}{\partial y}\left[\kappa_{n_f}(T)\frac{\partial T}{\partial y}\right]. \tag{3}$$

In the above system of equations, u and v represent velocities in the x and y directions, respectively; k is the permeability of the medium; T is the nanofluid temperature; $\mu_{n_f}(T)$ the nanofluid temperature-dependent viscosity; ρ_{n_f} the nanofluid density; $(C_p)_{n_f}$ is the specific heat at constant pressure; and $\kappa_{n_f}(T)$ is the thermal conductivity of the nanofluid.

The appropriate partial slip boundary conditions for velocity and temperature are:

$$u = L_1\frac{\partial u}{\partial y}, \quad v = V_w \quad \text{at} \quad y = 0; \quad u \to u_\infty \quad \text{as} \quad y \to \infty, \tag{4}$$

$$T = T_w + D_1\frac{\partial T}{\partial y} \quad \text{at} \quad y = 0; \quad T \to T_\infty \quad \text{as} \quad y \to \infty. \tag{5}$$

Here, $L_1 = LR_{e_x}$ is the velocity slip factor, and $D_1 = D\sqrt{LR_{e_x}}$ is the thermal slip factor with L and D the initial values of velocity and thermal slip factors; and $R_{e_x} = \frac{u_\infty x}{\nu_f}$ is the local Reynolds's number with $\nu_f = \frac{\mu_f}{\rho_f}$ the kinematic viscosity of the base fluid. V_w shows the mass transfer at the surface with $V_w > 0$ for injection and $V_w < 0$ for suction.

Following Maxwell [4], Bhaskar et al. [34] and Arunachalam [42], the nanofluid's physical parameters are taken as:

$$\rho_{n_f} = (1 - \phi\rho_f + \rho_s), \qquad (\rho C_p)_{n_f} = (1 - \phi(\rho C_p)_f + \phi(\rho C_p)_s), \tag{6}$$

$$\mu_{nf} = \mu^*_{n_f}\left[a + b(T_w - T)\right], \qquad \kappa_{nf}(T) = \kappa^*_{n_f}\left[1 + \epsilon\frac{T - T_\infty}{T_w - T_\infty}\right], \tag{7}$$

$$\mu^*_{n_f} = \mu_f(1 - \phi)^{-2.5}, \qquad \frac{\kappa^*_{nf}}{\kappa_f} = \frac{(\kappa_s + 2\kappa_f) - 2\phi(\kappa_f - \kappa_s)}{(\kappa_s + 2\kappa_f) + \phi(\kappa_f - \kappa_s)}. \tag{8}$$

In Equations (6)–(8), ϕ is the nanoparticle volume fraction coefficient, ρ_f the density of the base fluid, ρ_s the density of the nanoparticles, $(Cp)_f$ the specific heat capacity of the base fluid, $(Cp)_s$ the specific heat capacity of the nanoparticles, $\mu^*_{n_f}$ and $\kappa^*_{n_f}$ the constant values of the coefficient of viscosity and thermal conductivity of the nanofluid, respectively, and a, b and ϵ the constants with $b > 0$, μ_f, κ_f and κ_s the coefficient of viscosity, thermal conductivity of base fluid and nanoparticles, respectively.

3. Solution of the Problem

We introduce the relation for u, v and T as:

$$u = \frac{\partial\psi}{\partial y}, \qquad v = -\frac{\partial\psi}{\partial x}, \qquad \theta(\eta) = \frac{T - T_\infty}{T_w - T_\infty}, \tag{9}$$

where the stream function $\psi(\eta)$ and dimensionless similarity variable η are defined by (see, for example, Bhattacharyya et al. [43])

$$\psi = \nu_f R_{ex}^{\frac{1}{2}} f(\eta), \qquad \eta = \frac{y}{x}R_{ex}^{\frac{1}{2}}. \tag{10}$$

Equations (9) and (10) together with Equations (6)–(8) reduce the boundary value problem (2)–(5) to:

$$(a + A - A\theta)f''' + (1-\phi)^{2.5}(1 - \phi + \phi\frac{\rho_s}{\rho_f})(\frac{1}{2}ff'') - A\theta' f'' - k^*(a + A - A\theta)(f' - 1) = 0, \tag{11}$$

$$(1 + \epsilon\theta)\theta'' + \epsilon\theta'^2 + P_r(\frac{k_f}{k_{n_f}})\left(1 - \phi + \phi\frac{(\rho C_p)_s}{(\rho C_p)_f}\right)\left(\frac{1}{2}f\theta'\right) = 0, \tag{12}$$

$$f(\eta) = S, \quad f'(\eta) = \delta f''(\eta) \quad \text{at} \quad \eta = 0; \quad f'(\eta) \to 1 \quad \text{as} \quad \eta \to \infty \tag{13}$$

$$\theta(\eta) = 1 + \Delta\theta'(\eta) \quad \text{at} \quad \eta = 0; \quad \theta(\eta) \to 0 \quad \text{as} \quad \eta \to \infty, \tag{14}$$

where $A = b(T_w - T_\infty)$ is the viscosity parameter, $k^* = \frac{1}{D_{a_x} R_{e_x}}$ is the permeability parameter, $D_{a_x} = \frac{k}{x^2}$ is the local Darcy number, $P_r = \frac{\nu_f}{\alpha_f}$ is the Prandtl number, $\alpha_f = \frac{k_f}{(\rho C_p)_f}$ is the diffusivity parameter, ϵ is the thermal conductivity parameter, $\delta = \frac{L u_\infty}{\nu_f}$ is the velocity slip parameter and $\Delta = \frac{D u_\infty}{\nu_f}$ is the thermal slip parameter.

The important physical quantities of interest are the skin friction coefficient C_F (rate of shear stress) and the local Nusselt number N_{u_x} (rate of heat transfer at the surface). The skin friction coefficient and the Nusselt number are defined as:

$$C_F = \frac{\tau_w}{\rho U_w^2}, \qquad N_{u_x} = \frac{x q_w}{k_f (T_w - T_\infty)}, \tag{15}$$

where the local wall shear stress τ_w and the heat transfer from the plate q_w are given by:

$$\tau_w = -\mu_{n_f} (\frac{\partial u}{\partial y})_{y=0}, \qquad q_w = \kappa_{n_f} (\frac{\partial T}{\partial y})_{y=0} \tag{16}$$

with u the flow velocity parallel to the porous plate and y the distance to the plate. Using Equation (16), the dimensionless forms of Equation (15) become:

$$C_f R_{e_x}^{1/2} (1-\phi)^{2.5} = -f''(0), \qquad N_{u_x} R_{e_x}^{-1/2} (\frac{k_f}{k_{n_f}}) = -\theta'(0). \tag{17}$$

4. Numerical Method for Solution

The nonlinear coupled ordinary differential Equations (11) and (12) subject to boundary conditions (13) and (14) form the two-point boundary value problem and are solved numerically using the shooting method. In order to use the shooting method, first we convert (11) and (12) to a system of first order differential equations:

$$f' = p, \quad p' = q, \quad \theta' = z, \tag{18}$$

$$q' = \frac{1}{(a + A - A\theta)} \left[(1-\phi)^{2.5} (1 - \phi + \phi \frac{\rho_s}{\rho_f}) (-\frac{1}{2} f q) + A z q + k^* (a + A - A\theta)(p - 1) \right], \tag{19}$$

$$z' = \frac{1}{(1 + \epsilon\theta)} \left[-\epsilon z^2 - P_r (\frac{k_f}{k_{n_f}}) \left(1 - \phi + \phi \frac{(\rho C_p)_s}{(\rho C_p)_f} \right) \left(\frac{1}{2} f z \right) \right]. \tag{20}$$

The boundary conditions become:

$$f(0) = S, \quad p(0) = 1 + \delta q(0), \quad \theta(0) = 1 + \Delta z(0). \tag{21}$$

In order to solve the initial value problem (19)–(21) with the shooting method, we require an initial guess for $q(0)$ and $z(0)$. The required values of $q(0)$ and $z(0)$ are chosen randomly, and numerical solutions are obtained using fourth order Runge–Kutta method. The numerical values for $q(0)$ and $z(0)$ are adjusted using Newton's method to give better approximation to the solution. The step size is taken as 0.01, and the process is repeated until the solutions achieve the accuracy of 10^{-6}. To ensure the numerical accuracy, we have compared our results with the results of Bhattacharyya et al. [43] for velocity and temperature profiles with $A = \epsilon = \phi = 0, a = 1, S = 0.2, \delta = \Delta = 0.1$ and $P_r = 0.3$. The comparison is shown in Figure 2 and is found to be in excellent agreement.

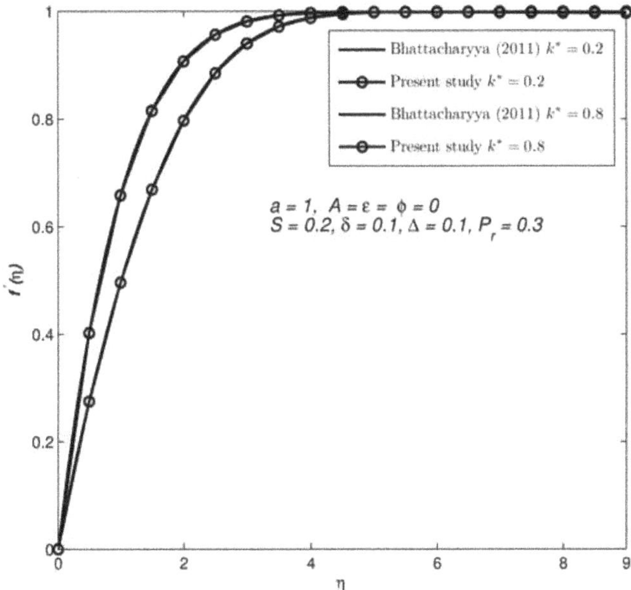

Figure 2. Comparison of the results for the velocity $f'(\eta)$ profiles for different values of permeability parameter k^*, with Bhattachaaya et al. [43].

The thermophysical properties of the base fluid and nanoparticles are given in Table 1

Table 1. Thermophysical properties of base fluid and nanoparticles [44].

Physical Properties	Units	Base Fluid Water	Nanoparticles (Solid) Cu (300-K)
Density ρ	(kg/m^3)	997.1	8933
Specific heat Cp	(J/kg·K)	4179	385
Thermal conductivity κ	(W/m·K)	0.613	401

5. Numerical Results and Discussion

In this section, the numerical results calculated for the velocity and temperature profiles are presented through graphs and tables. The computations are performed to study the effects of the variation of permeability parameter k^*, nanofluid volume concentration parameter ϕ, velocity slip parameter δ, thermal slip parameter Δ, suction and injection parameter S, viscosity parameter A and variable thermal conductivity ϵ on the velocity and temperature profiles of the Cu-water nanofluid. The behavior of the skin friction coefficient and Nusselt number with the variation in physical parameters is also shown in Table 2.

Table 2. Values of skin friction $= -f''(0)$ and Nusselt number $= -\theta'(0)$.

κ^*	A	ϕ	ϵ	S	δ	Δ	$-f''(0)$	$-\theta'(0)$
$a = 1$								
0.2	0.2	0.2	0.3	0.2	0.1	0.1	0.5844	0.2513
0.8							0.9158	0.2678
0.1	0.2	0.2	0.3	0.1	0.1	0.1	0.6139	0.6594
0.4							0.8145	0.6951
0.9							1.0470	0.7290
0.4	0.1	0.2	0.3	0.1	0.1	0.1	0.7858	0.7996
	0.6						0.8780	0.8098
	1.5						1.0186	0.8239
0.4	0.2	0.0	0.3	0.1	0.1	0.1	0.7815	0.9284
		0.05					0.7880	0.8621
		0.2					0.8145	0.6951
0.4	0.2	0.2	0.1	0.1	0.1	0.1	0.8158	0.7709
			0.4				0.8139	0.6639
			0.8				0.8115	0.5693
0.4	0.2	0.2	0.3	0.1	0.1	0.1	0.7742	0.6135
				0.2			0.8145	0.6951
				0.3			0.8553	0.7794
0.4	0.2	0.2	0.3	0.2	0.1	0.1	0.8145	0.6951
					0.3		0.7163	0.7404
					0.6		0.6007	0.7869
0.4	0.2	0.2	0.3	0.2	0.1	0.1	0.8145	0.6951
						0.3	0.8052	0.6188
						0.6	0.7949	0.5292

The influence of the permeability parameter k^* on the velocity and temperature profiles in the presence of slip at the boundary is depicted in Figures 3 and 4. The velocity and temperature profiles are plotted for several values of permeability parameter k^* for the Cu-water nanofluid. It is observed that the velocity of the nanofluid increases with the increase in the permeability of the medium and consequently decreases the thickness of the momentum boundary layer. This is due to the fact that the increase in permeability reduces the magnitude of the Darcian body force (inversely proportional to the permeability) and enhances the motion of the fluid in the boundary layer. In other words, progressively less drag is experienced by the flow, and flow retardation thereby decreases. From Figure 4, it is noticed that the temperature $\theta(\eta)$ at a fixed distance from the plate decreases with the increase in k^*. The permeability parameter is inversely proportional to the density of the base fluid, hence the increase in k^* causes a decrease in the density and temperature of the nanofluid within the boundary layer. In conclusion, the increase in the permeability of the porous medium decreases the thickness of momentum and thermal boundary layers and eventually increases the heat transfer rate.

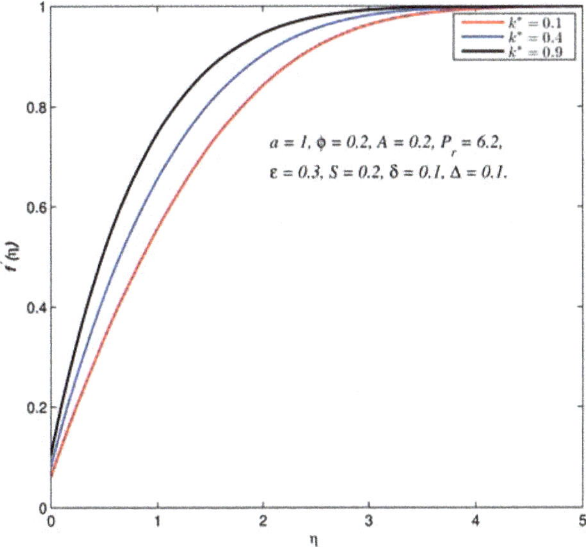

Figure 3. Velocity $f'(\eta)$ profiles for different values of permeability parameter k^*.

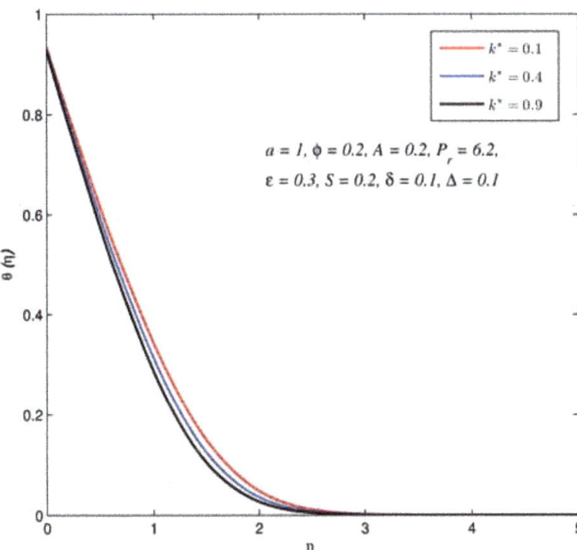

Figure 4. Temperature $\theta(\eta)$ profiles for different values of permeability parameter k^*.

The effect of nanoparticle volume concentration parameter ϕ on the velocity and temperature profiles of the Cu-water nanofluid is shown in Figures 5 and 6. It is observed that the velocity of the fluid decreases, whereas the temperature of the nanofluid increases with the increase in volume concentration parameter ϕ. This illustrates the agreement with the physical behavior of the nanofluids, i.e., the increase in the volume of nanoparticles causes an increase in the thermal conductivity of the fluid, which leads to the increase in the thickness of the thermal boundary layer.

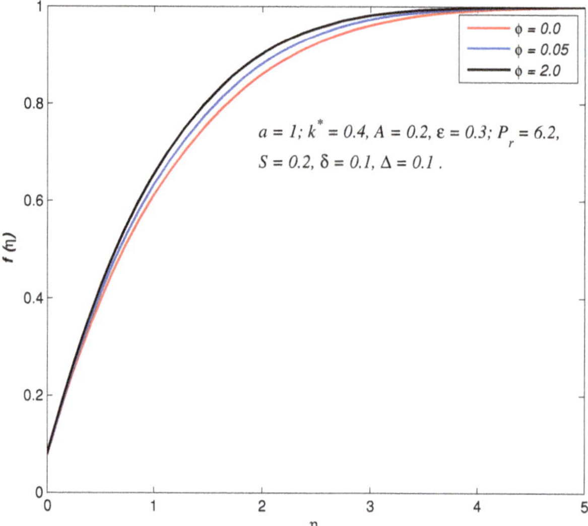

Figure 5. Velocity $f'(\eta)$ profiles for different values of volume fraction coefficient ϕ.

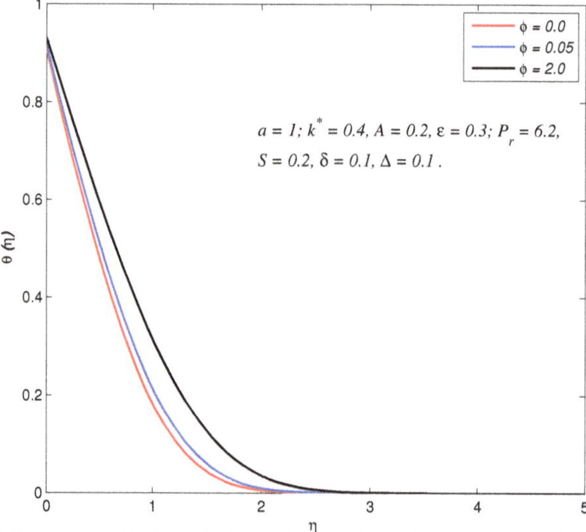

Figure 6. Temperature $\theta(\eta)$ profiles for different values of volume fraction coefficient ϕ.

Figures 7 and 8, respectively, show the nanofluid velocity and temperature profiles for different values of viscosity parameter A. The comparison of curves in Figure 7 shows that the velocity of the nanofluid initially increases with the increase in viscosity parameter A. This increase in velocity corresponds to a reduction in the thickness of the momentum boundary layer. Moreover, the cross-over point is also observed for velocity profiles in Figure 7. The velocity profiles exhibit opposite behaviors after crossing the cross-over point, that is the velocity decreases with the increasing values of viscosity parameter A after the crossing-over point. This corresponds to an increase in the thickness of the boundary layer. It is observed from Figure 8 that the increase in the viscosity parameter enhances the heat transfer rate and decreases the thickness of the thermal boundary layer. The impact of A on the velocity profiles is more pronounced than on the temperature profiles.

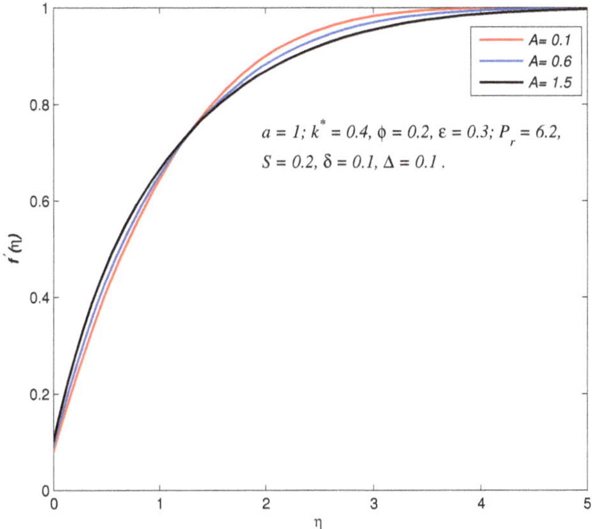

Figure 7. Velocity $f'(\eta)$ profiles for different values of viscosity parameter A.

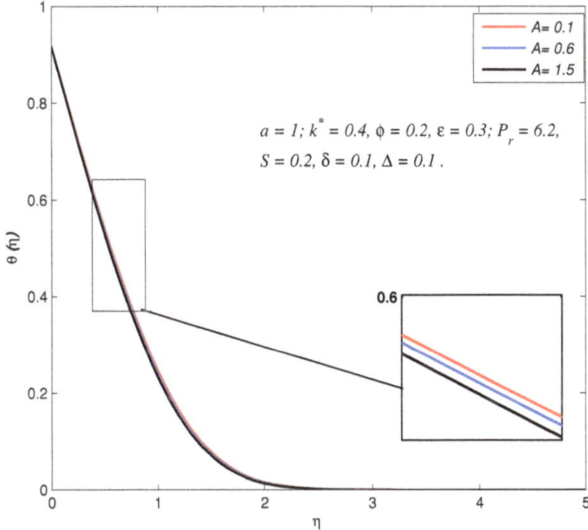

Figure 8. Temperature $\theta(\eta)$ profiles for different values of viscosity parameter A.

Figures 9 and 10 depict the velocity and temperature profiles for the specified values of thermal conductivity parameter ϵ. It is noticed that the variation in ϵ greatly affects the temperature profiles, as compared to the velocity profiles. The variation in velocity profiles show an opposite effect as the variation in viscosity parameter A, i.e., the increase in thermal conductivity initially causes the decrease in fluid velocity and shows the opposite behavior after the cross-over point; whereas an increase in ϵ results in an increase in thermal conductivity, thereby raising the fluid temperature across the boundary layer. It would also increase the thermal boundary layer thickness.

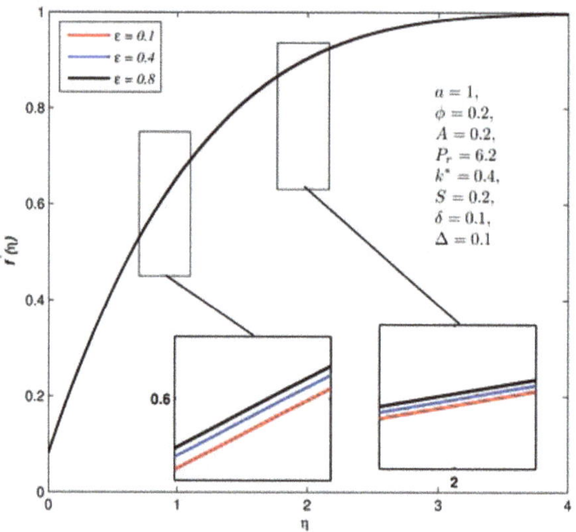

Figure 9. Velocity $f'(\eta)$ profiles for different values of thermal conductivity parameter ϵ.

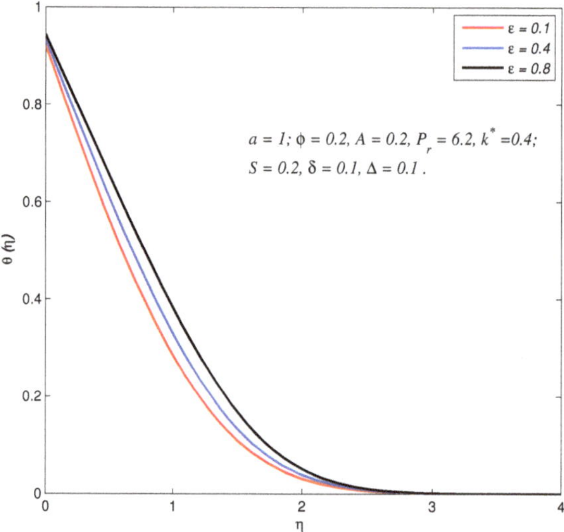

Figure 10. Temperature $\theta(\eta)$ profiles for different values of ϵ.

In Figure 11, the effect of velocity slip parameter δ on the velocity profile of the Cu-water nanofluid is presented. The comparison of the curves shows that the increase in the velocity slip at the boundary increases the fluid velocity within the boundary layer. This is due to the positive value of the fluid velocity adjacent to the surface of the plate and results in a reduction of the momentum boundary layer thickness. Moreover, the increase in magnitude of the slip parameter allows more fluid to slip past the plate, and accordingly, the flow through the boundary layer will increase. The temperature profiles in Figure 12 show the decrease in temperature and thermal boundary layer thickness with an increase in velocity slip parameter δ.

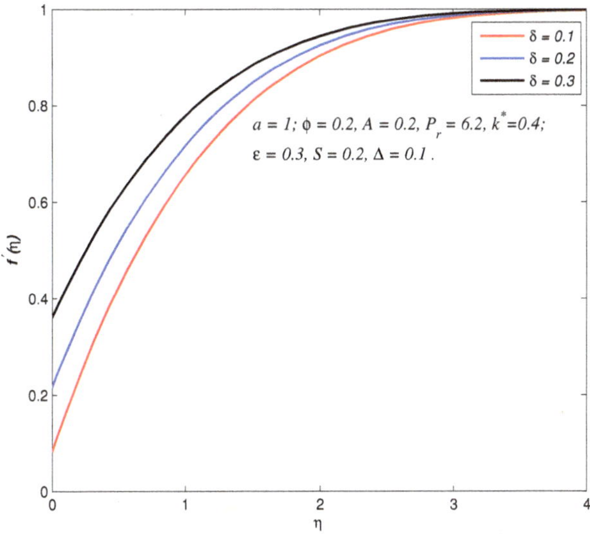

Figure 11. Velocity $f'(\eta)$ profiles for different values of velocity slip parameter δ.

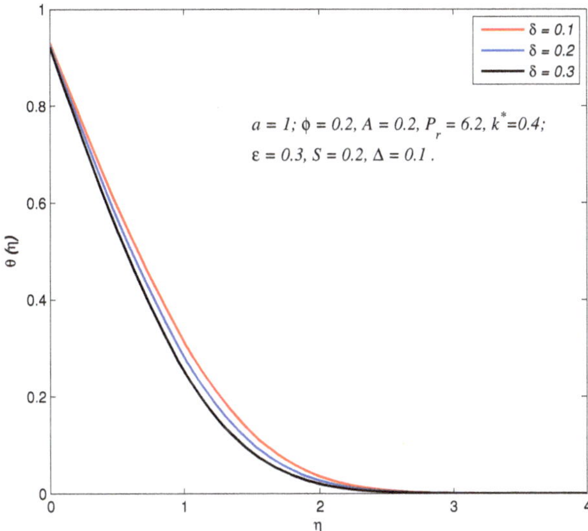

Figure 12. Temperature $\theta(\eta)$ profiles for different values of velocity slip parameter δ.

Figure 13 depicts the decrease in the thickness of the thermal boundary layer with the increase in thermal slip parameter Δ. This is because the increase in the thermal slip parameter causes less transfer of heat from the sheet to the fluid, which leads to a decrease in the boundary layer temperature.

Moreover, the momentum equation is dependent on $\theta(\eta)$, but no significant effect of thermal slip parameter Δ on the velocity profiles is noticed.

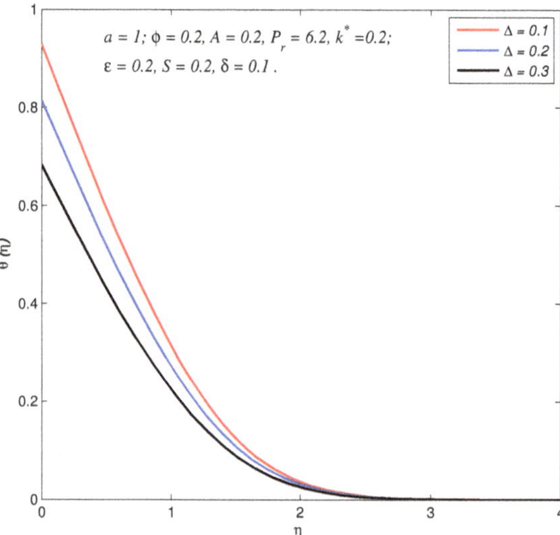

Figure 13. Temperature $\theta(\eta)$ profiles for different values of thermal slip parameter Δ.

The behavior of the velocity and temperature distribution for variation of suction $(S > 0)$ and blowing $(S < 0)$ parameter in the presence of slip conditions at the boundary are plotted in Figures 14–17. For $S > 0$, the fluid velocity increases as the fluid particles are sucked in the porous wall, which in turn reduces the thickness of the momentum boundary layer. On the other hand, for the case of blowing, i.e., $S < 0$, the opposite trend is observed. When suction $S > 0$ is increased, it refers to bringing the fluid close to the wall. This causes a decrease in the temperature profile and also decreases the thermal boundary layer. This entire phenomenon causes an increase in the rate of heat transfer. An opposite trend can be seen for the case of blowing $S < 0$.

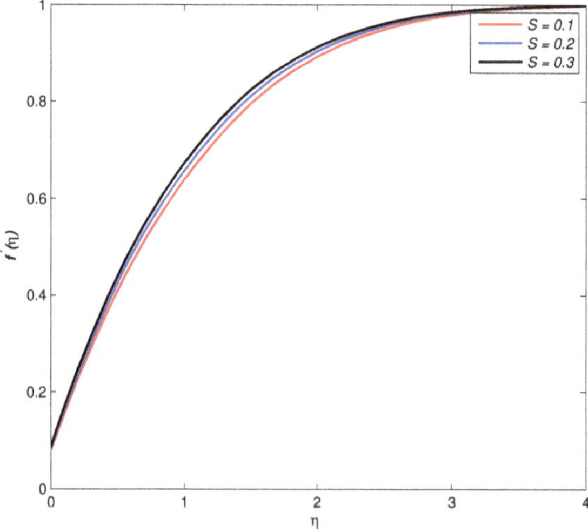

Figure 14. Velocity $f'(\eta)$ profiles for different values of suction parameter $S > 0$.

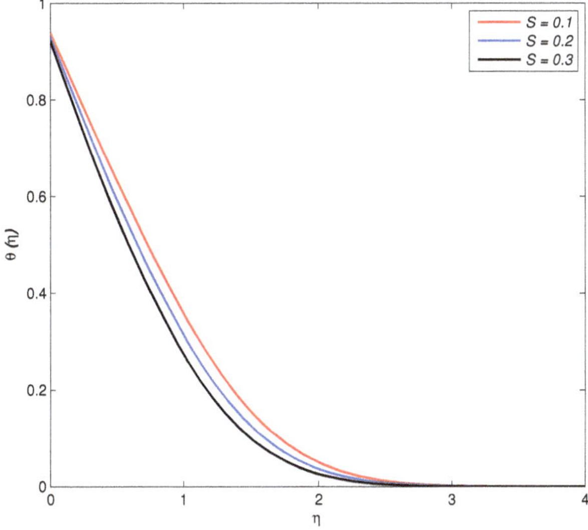

Figure 15. Temperature $\theta(\eta)$ profiles for different values of suction parameter $S > 0$.

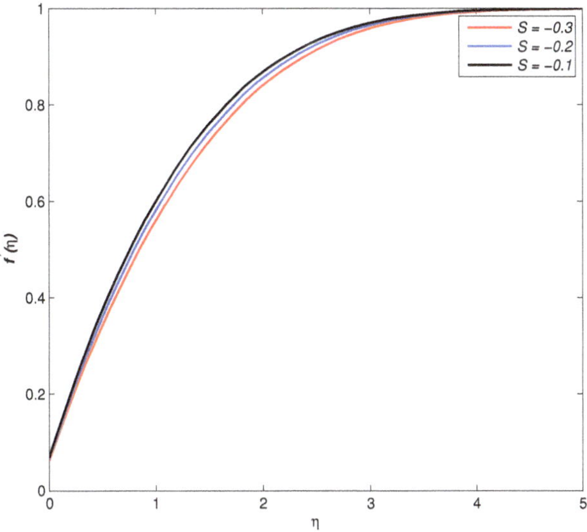

Figure 16. Velocity $f'(\eta)$ profiles for different values of blowing parameter $S < 0$.

The behavior of the skin friction coefficient and Nusselt number with the variation in different thermophysical parameters is shown in Table 2. It is evident that the skin friction coefficient increases with increasing values of permeability parameter k^*, viscosity parameter A, nanofluid volume concentration parameter ϕ and suction parameter S; whereas a decreasing trend is observed for increasing values of thermal conductivity parameter ϵ, velocity slip parameter δ and thermal slip parameter Δ. The increasing values of the skin friction coefficient correspond to the thinning of the velocity boundary layer; whereas the decreasing values of the skin friction coefficient correspond to fluid velocity at the surface approaching the free stream velocity. The negative value of the temperature gradient at the plate $-\theta'(0)$ is proportional to the rate of heat transfer at the surface of the plate. The rate of heat transfer at the surface is increasing for increasing values of permeability parameter k^*, viscosity parameter A, suction parameter S and velocity slip parameter δ. The rate of heat transfer decreases with the increase in the volume concentration parameter ϕ, the thermal conductivity parameter ϵ and the thermal slip parameter Δ.

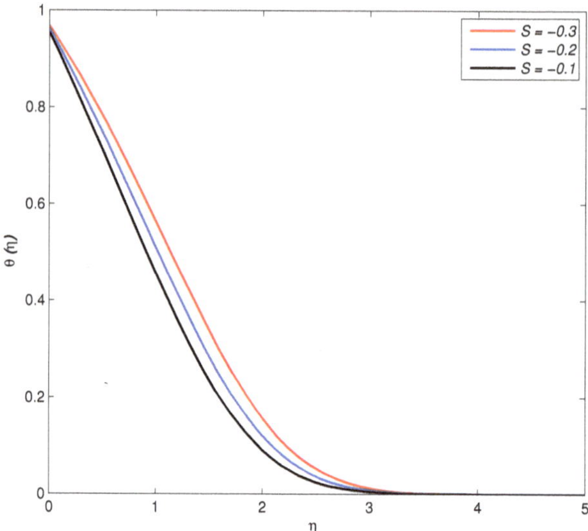

Figure 17. Temperature $\theta(\eta)$ profiles for different values of blowing parameter $S < 0$.

6. Conclusions

In the present research, we investigated the slip effects on the steady flow and heat transfer of nanofluids over a porous sheet embedded in a Darcy-type porous medium. The viscosity and the thermal conductivity of the nanofluids were considered as linear functions of temperature, and wall slip conditions were employed in terms of shear stress. The governing system of equations was reduced to the ordinary differential equations by suitable similarity transformations, and the reduced system was solved numerically using the shooting method. The influence of key thermophysical parameters on the velocity and temperature profiles, as well as on the skin friction coefficient and Nusselt number were presented and discussed through graphs and tables. The present model exploited a number of simplifications in order to focus on the principal effects of permeability, variable viscosity, variable thermal conductivity, nanofluid volume concentration and slip parameters. In our work, we showed that the increase in the permeability of the porous medium, the viscosity of the nanofluids and the velocity slip parameter decreased the momentum and thermal boundary layer thickness and eventually increased the rate of heat transfer; whereas the opposite trend was observed for the increase in the thermal conductivity parameter. The present simplified model can be generalized to reveal the effects of effective dynamic viscosity on the slip flow and heat transfer of nanofluids for the viscosity models proposed by Abu-Nada [45], Khanafer and Vafai [46] and Corcione [47]. Moreover, the analysis can be extended to include the results for different water-based nanofluids, and a comparison can be generated on the heat transfer characteristics of different nanofluids. Clearly, there is an opportunity to consider/extend this problem with non-Newtonian nanofluid models and to perform experimental work on these systems.

Author Contributions: A. A., S. H. and T. A. contributed in the modeling and wrote the paper. S.H. performed the simulations and processed all figures. C.M.K. contributed to the idea and the organization of the research work.

Conflicts of Interest: The authors declare no conflicts of interest.

References

1. Kreith, F.; Boehm, R.F. *Heat and Mass Transfer (Mechanical Engineering Handbook)*; CRC Press: Boca Raton, FL, USA, 1999.
2. Nield, D.; Bejan, A. *Convection in Porous Media*, 3rd ed.; Springer: New York, NY, USA, 2006.
3. Webb, R.; Kim, N. *Principles of Enhanced Heat Transfer*, 2nd ed.; John Wiley: New York, NY, USA, 1994.
4. Maxwell, J. *A Treatise on Electricity and Magnetism*, 2nd ed.; Clarendon Press: Oxford, UK, 1881.
5. Vafai, K. *Handbook of Porous Media*; Marcel Dekker: New York, NY, USA, 2000.
6. Ingham, D.; Pop, I. *Transport Phenomena in Porous Media*; Elsevier: Amsterdam, The Netherlands, 1998.
7. Choi, S.U.S. Enhancing thermal conductivity of fluids with nanoparticles. *ASME Int. Mech. Eng. Congr. Expo.* **1995**, *66*, 99–105.
8. Wang, X.; Xu, X.; Choi, S.U.S. Thermal conductivity of nanoparticle-fluid mixture. *J. Thermophys. Heat Transf.* **1999**, *13*, 474–480.
9. Keblinski, P.; Phillpot, S.R.; Choi, S.U.S.; Eastman, J.A. Mechanisms of heat flow in suspension of nano-sized particles (nanofluids). *Int. J. Heat Mass Transf.* **2002**, *45*, 855–863.
10. Buongiorno, J. Convective transport in nanofluids. *ASME J. Heat Transf.* **2006**, *128*, 240–250.
11. Keblinski, P.; Eastman, J.; Cahill, D. Nanofluids for thermal transport. *Mater. Today* **2005**, *8*, 36–44.
12. Wang, X.Q.; Mujumdar, A.S. Heat transfer characteristics of nanofluids: A review. *Int. J. Therm. Sci.* **2007**, *46*, 1–19.
13. Nield, D.; Kuznetsov, A. Thermal instability in a porous medium layer saturated by a nanofluid. *Int. J. Heat Mass Transf.* **2009**, *52*, 5796–5801.
14. Kuznetsov, A.V.; Nield, D.A. Natural convective boundary-layer flow of a nanofluid past a vertical plate. *Int. J. Therm. Sci.* **2010**, *49*, 243–247.
15. Sun, Q.; Pop, I. Free convection in a triangle cavity filled with a porous medium saturated with nanofluids with flush mounted heater on the wall. *Int. J. Therm. Sci.* **2011**, *50*, 2141–2153.
16. Khan, W.; Aziz, A. Double-diffusive natural convective boundary layer flow in a porous medium saturated with a nanofluid over a vertical plate: Prescribed surface heat, solute and nanoparticle fluxes. *Int. J. Therm. Sci.* **2011**, *50*, 2154–2160.
17. Yadav, D.; Bhargava, R.; Agrawal, G.S. Boundary and internal heat source effects on the on set of darcy-brinkman convection in a porous layer saturated by nanofluid. *Int. J. Therm. Sci.* **2012**, *60*, 244–254.
18. Uddin, M.J.; Khan, W.A.; Ismail, A. Free convection of non-newtonian nano-fluids in porous media with gyrotactic micro organisms. *Transp. Porous Media* **2013**, *97*, 241–252.
19. Servati, V.; Javaherdeh, K.; Ashorynejad, H.R. Magnetic field effects on force convection flow of a nanofluid in a channel partially filled with porous media using lattice boltzmann method. *Adv. Powder Technol.* **2014**, *25*, 666–675.
20. Ahmad, S.; Pop, I. Mixed convection boundary layer flow from a vertical flat plate embedded in a porous medium filled with nanofluids. *Int. Commun. Heat Mass Transf.* **2010**, *37*, 987–991.
21. Cimpean, D.S.; Pop, I. Fully developed mixed convection flow of a nanofluid through an inclined channel filled with a porous medium. *Int. J. Heat Mass Transf.* **2012**, *55*, 907–914.
22. Mahdi, R.A.; Mohammed, H.A.; Munisamy, K.M.; Saeid, N.H. Review of convection heat transfer and fuid fow in porous media with nanofuid. *Renew. Sustain. Energy Rev.* **2005**, *41*, 715–734.
23. Uphill, S.J.; Cosgrove, T.; Briscoe, W.H. Flow of nanofluids through porous media: Preserving timber with colloid science. *Colloids Surf. A Physicochem. Eng. Asp.* **2014**, *460*, 38–50.
24. Zhang, C.; Zheng, L.; Zhang, X.; Chen, G. MHD flow and radiation heat transfer of nanofluids in porous media with variable surface heat flux and chemical reaction. *Appl. Math. Model.* **2015**, *39*, 165–181.
25. Sheikholeslami, M.; Ellahi, R. Electrohydrodynamic Nanofluid Hydrothermal Treatment in an Enclosure with Sinusoidal Upper Wall. *Appl. Sci.* **2015**, *5*, 294–306.
26. Reddy, J.V.R.; Sugunamma, V.; Sandeep, N.; Sulochana, C. Influence of chemical reaction, radiation and rotation on MHD nanofluid flow past a permeable flat plate in porous medium. *J. Niger. Math. Soc.* **2016**, *35*, 48–65.

27. Luk, J.; Mutharasan, R.; Apelian, D. Experimental observations of wall slip: Tube and packed bed flow. *Ind. Eng. Chem. Res.* **1987**, *26*, 1609–1616.
28. Kalyon, D.M. Apparent slip and viscoplasticity of concentrated suspensions. *J. Rheol.* **2005**, *49*, 621–640.
29. Khan, W.A.; Khan, Z.H.; Rahi, M. Fluid flow and heat transfer of carbon nanotubes along a flat plate with navier slip boundary. *Appl. Nanosci.* **2014**, *4*, 633–641.
30. Zheng, L.; Niu, J.; Zhang, X.; Gao, Y. MHD Flow and Heat Transfer over a Porous Shrinking Surface with Velocity Slip and Temperature Jump. *Math. Comput. Model.* **2012**, *56*, 133–144.
31. Zheng, L.; Zhang, C.; Zhang, X.; Zhang, J. Flow and radiation heat transfer of a nanofluid over a stretching sheet with velocity slip and temperature jump in porous medium. *J. Frankl. Inst.* **2013**, *350*, 990–1007.
32. Uddin, M.J.; Khan, W.A.; Amin, N.S. g-jitter mixed convective slip flow of nanofluid past a permeable stretching sheet embedded in a darcian porous media with variable viscosity. *PLoS ONE* **2014**, *9*, e99384.
33. Noghrehabadi, A.; Pourrajab, R.; Ghalambaz, M. Effect of partial slip boundary condition on the flow and heat transfer of nanofluids past stretching sheet prescribed constant wall temperature. *Int. J. Therm. Sci.* **2012**, *54*, 253–261.
34. Bhaskar, N.R.; Poornima, T.; Sreenivasulu, P. Influence of variable thermal conductivity on MHD boundary layer slip flow of ethylene-glycol based Cu nanofluids over a stretching sheet with convective boundary condition. *Int. J. Eng. Math.* **2014**, *2014*, 905158.
35. Noghrehabadia, A.; Ghalambaza, M.; Ghanbarzadeha, A. Effects of variable viscosity and thermal conductivity on natural-convection of nanofluids past a vertical plate in porous media. *J. Mech.* **2014**, *30*, 265–275.
36. Uddin, M.J.; Pop, I.; Ismail, A.I.M. Free convection boundary layer flow of a nanofluid from a convectively heated vertical plate with linear momentum slip boundary condition. *Sains Malays.* **2012**, *4*, 1475–1482.
37. Ibrahim, W.; Shankar, B. MHD boundary layer flow and heat transfer of a nanofluid past a permeable stretching sheet with velocity, thermal and solutal slip boundary conditions. *Comput. Fluids* **2013**, *75*, 1–10.
38. Ellahi, R. The effects of MHD and temperature dependent viscosity on the flow of non-Newtonian nanofluid in a pipe: analytical solutions. *Appl. Math. Model.* **2013**, *37*, 1451–1467.
39. Sharma, R.; Ishak, A. Second order slip flow of cu-water nanofluid over a stretching sheet with heat transfer. *WSEAS Trans. Fluid Mech.* **2014**, *9*, 26–33.
40. Mohyuddin, S.T.; Khan, U.; Ahmed, N.; Hassan, S.M. Magnetohydrodynamic Flow and Heat Transfer of Nanofluids in Stretchable Convergent/Divergent Channels. *Appl. Sci.* **2015**, *5*, 1639–1664.
41. Pandey, A.K.; Kumar, M. Effect of viscous dissipation and suction/injection on MHD nanofluid flow over a wedge with porous medium and slip. *Alex. Eng. J.* **2016**, doi:10.1016/j.aej.2016.08.018.
42. Arunachalam, M.; Rajappa, N.R. Forced convection in liquid metals with variable thermal conductivity and capacity. *Acta Mech.* **1978**, *31*, 25–31.
43. Bhattacharyya, K.; Layek, G.C.; Gorla, R.S.R. Boundary layer slip flow and heat transfer past a stretching sheet with temperature dependent viscosity. *Therm. Energy Power Eng.* **2013**, *2*, 38–43.
44. Shankar, B.; Yirga, Y. Unsteady heat and mass transfer in mhd flow of nanofluids over stretching sheet with a non-uniform heat source/sink. *World Acad. Sci. Eng. Technol. Int. J. Math. Comput. Stat. Nat. Phys. Eng.* **2013**, *7*, 12.
45. Abu-Nada, E. Effects of variable viscosity and thermal conductivity of Al_2O_3–water nanofluid on heat transfer enhancement in natural convection. *Int. J. Heat Fluid Flow* **2009**, *30*, 679–690.
46. Khanafer, K.; Vafai, K. A critical synthesis of thermophysical characteristics of nanofluids. *Int. J. Heat Mass Transf.* **2011**, *54*, 4410–4428.
47. Corcione, M. Empirical correlating equations for predicting the effective thermal conductivity and dynamic viscosity of nanofluids. *Energy Convers. Manag.* **2011**, *52*, 789–793.

Article

Natural Convection Flow of Fractional Nanofluids Over an Isothermal Vertical Plate with Thermal Radiation

Constantin Fetecau [1,*,†], Dumitru Vieru [2,†] and Waqas Ali Azhar [3,†]

1 Academy of Romanian Scientists, Bucharest 050094, Romania
2 Department of Theoretical Mechanics, Technical University of Iasi, Iasi 700050, Romania; dumitru_vieru@yahoo.com
3 Abdus Salam School of Mathematical Sciences, Government College University, Lahore 54000, Pakistan; waqasaliazhar@gmail.com
* Correspondence: c_fetecau@yahoo.com; Tel.: +40-072-165-6339
† These authors contributed equally to this work.

Academic Editor: Yulong Ding
Received: 27 November 2016; Accepted: 1 March 2017; Published: 3 March 2017

Abstract: The studies of classical nanofluids are restricted to models described by partial differential equations of integer order, and the memory effects are ignored. Fractional nanofluids, modeled by differential equations with Caputo time derivatives, are able to describe the influence of memory on the nanofluid behavior. In the present paper, heat and mass transfer characteristics of two water-based fractional nanofluids, containing nanoparticles of CuO and Ag, over an infinite vertical plate with a uniform temperature and thermal radiation, are analytically and graphically studied. Closed form solutions are determined for the dimensionless temperature and velocity fields, and the corresponding Nusselt number and skin friction coefficient. These solutions, presented in equivalent forms in terms of the Wright function or its fractional derivatives, have also been reduced to the known solutions of ordinary nanofluids. The influence of the fractional parameter on the temperature, velocity, Nusselt number, and skin friction coefficient, is graphically underlined and discussed. The enhancement of heat transfer in the natural convection flows is lower for fractional nanofluids, in comparison to ordinary nanofluids. In both cases, the fluid temperature increases for increasing values of the nanoparticle volume fraction.

Keywords: free convection; thermal radiation; fractional nanofluids; exact solutions

1. Introduction

Natural convection flows have been extensively studied due to their multiple engineering applications. Such flows over an infinite plate are usually met in different engineering processes, including petroleum resource gas production, and geothermal reservoirs, thermal insulation, etc. (see [1–3]). The effects of thermal radiation are also important in geophysics, and geothermic, chemical, and ceramics processing, and they have been investigated by many researchers. A short presentation of the main results, up until 2007, is given by Ghosh and Beg [4], who studied the convective radiative heat transfer over a hot vertical surface in porous medium. Moreover, the effects of thermal radiation on nanofluid flows have been studied by many scholars. Mondal et al. [5] considered the unsteady magneto-hydrodynamic axi-symmetric stagnation-point flow over a shrinking sheet with Navier slip, and the temperature-dependent thermal conductivity. Magneto-hydrodynamic (MHD) flows of nanofluids, with radiation heat transfer over a flat plate with a variable heat flux and first-order chemical reaction, were studied by Zhang et al. [6]. A numerical study on Cu-water and Ag-water nanofluids, focusing on the radiation effects over a stretching sheet, was made by Abd Elazem et al. [7].

Nowadays, it is well known that one way of enhancing the thermal conductivity of fluids is by suspending the metallic particles, such as alumina, gold, copper, iron, or titanium, in fluids [8]. These particles, also called nanoparticles, have a diameter of less than 100 nm and the obtained solution is named nanofluid. The concept of a nanofluid seems to have been introduced by Choi [9], and based on his work, many researchers have focused their attention on the heat transfer in natural convection flows of nanofluids. Khan and Aziz [10] have studied the natural convection flow of a nanofluid over a vertical plate with a uniform surface heat flux. Turkyilmazoglu [11] provided exact solutions for the MHD slip flow of a nanofluid over a stretching/shrinking sheet, while Bachok et al. [12] studied the heat transfer characteristics on a moving plate in a nanofluid. Interesting exact solutions have also been obtained by Turkyilmazoglu and Pop [13], for the velocity and temperature fields corresponding to the natural convection flow of some nanofluids, past an infinite vertical plate with radiation effects. Radiation and magnetic effects on the natural convection flow of a nanofluid, past an infinite vertical plate with a heat source, have been studied by Mohankrishna et al. [14]. Ellahi [15] studied the effects of MHD and temperature-dependent viscosity on the flow of a non-Newtonian nanofluid in a pipe, while an analysis of the flow and heat transfer of water and ethylene glycol-based Cu nanoparticles between two parallel disks with suction/injection effects, has been provided by Rizwan Ul Haq et al. [16]. Of course, the list of such studies can continue, but we close it with some of the most interesting analytical and numerical results that have been obtained in [17–27].

However, none of these papers took into consideration the fractional derivatives in their governing equations, although the fractional models have been found to be quite flexible in describing the complex behavior of many materials. More recently, it seems that fractional partial differential equations may be used to describe some physical phenomena more accurately, when compared to the corresponding partial differential equations. Our interest here is to provide exact solutions for the temperature and velocity fields corresponding to the radiative natural convection flow of fractional nanofluids over an infinite vertical plate with heat and mass transfer, and to investigate the enhancement of heat transfer in such a flow, utilizing the fractional model. The associated skin friction coefficient and Nusselt number will be also determined. These solutions, which are presented in equivalent forms in terms of the Wright functions or its fractional derivatives, are reduced to similar solutions, corresponding to ordinary nanofluids [13]. Finally, the influence of the fractional parameter on the thermal and hydrodynamic response of physical interest, is graphically underlined and discussed.

2. Statement of the Problem

Let us consider the unsteady free convection flow and heat transfer of a nanofluid, modeled by the Caputo time-fractional derivative, past an infinite vertical plate situated in the (x_1, z_1)-plane of a fixed Cartesian coordinate system $Ox_1y_1z_1$. At the initial moment $t_1 = 0$, the fluid and the plate are at rest, with a constant ambient temperature T_∞. We also consider the radiation effect and assume the radiative heat flux to be applied, perpendicular to the plate. Since the plate is infinite, all of the physical quantities describing the fluid motion are functions of y_1 and t_1. The fluid is a water-based nanofluid containing nano particles of CuO or Ag, whose thermo-physical properties are given in Table 1 [13,28].

Table 1. Thermo-physical properties of water and nanoparticles.

Basic Fluid/Nanoparticles	ρ (Kg/m^3)	C_p(J/Kg K)	k (W/m·K)	β × 10^5 (1/K)
Pure water	997.1	4179	0.613	21
Copper oxide (CuO)	6320	531.8	76.5	1.80
Silver (Ag)	10,500	235	429	1.89

Assuming a small difference between the fluid temperature $T(y_1, t_1)$ and the stream temperature T_∞, and adopting the Rosseland approximation [5], the radiative heat flux $q_r(y_1, t_1)$ can be linearized to:

$$q_r(y_1, t_1) = -\frac{16\sigma^* T_\infty{}^3}{3k^*}\frac{\partial T(y_1, t_1)}{\partial y_1}, \quad (1)$$

where σ^* is the Stefan-Boltzman constant and k^* is the mean absorption coefficient.

In the following, we consider the nanofluid model proposed by Tiwari and Das [29], and take into consideration the usual Boussinesq's approximation. In this case, the governing equations can be written as [13]:

$$\rho_{nf}\frac{\partial u_1(y_1, t_1)}{\partial t_1} = \mu_{nf}\frac{\partial^2 u_1(y_1, t_1)}{\partial y_1{}^2} + g(\rho\beta)_{nf}[T(y_1, t_1) - T_\infty]; y_1, t_1 > 0, \quad (2)$$

$$(\rho c_p)_{nf}\frac{\partial T(y_1, t_1)}{\partial t_1} = k_{nf}(1 + \frac{16\sigma^* T_\infty{}^3}{3k_{nf}k^*})\frac{\partial^2 T(y_1, t_1)}{\partial y_1{}^2}; y_1, t_1 > 0. \quad (3)$$

If no slipping exists between the fluid and the plate, the appropriate initial and boundary conditions are:

$$u_1(y_1, 0) = 0,\ T(y_1, 0) = T_\infty;\ y_1 \geq 0, \quad (4)$$

$$u_1(0, t_1) = 0,\ T(0, t_1) = T_w;\ t_1 > 0, \quad (5)$$

$$u_1(y_1, t_1) \to 0,\ T(y_1, t_1) \to 0 \text{ as } y_1 \to \infty. \quad (6)$$

In the above relations, $u_1(y_1, t_1)$ is the fluid velocity in the x_1-vertical direction, T_w is the constant plate temperature ($T_w > T_\infty$ or $T_w < T_\infty$ corresponds to the heated or cooled plate, respectively), g is the acceleration due to gravity, ρ_{nf} is the density of the nanofluid, μ_{nf} is the dynamic viscosity of the nanofluid, and β_{nf} is the thermal expansion coefficient of the nanofluid. Their expressions, as well as the expression of $(\rho c_p)_{nf}$, are given by:

$$\rho_{nf} = (1-\varphi)\rho_f + \varphi\rho_s,\ \mu_{nf} = \frac{\mu_f}{(1-\varphi)^{2.5}}, \quad (7)$$

$$(\rho\beta)_{nf} = (1-\varphi)(\rho\beta)_f + \varphi(\rho\beta)_s,\ (\rho c_p)_{nf} = (1-\varphi)(\rho c_p)_f + \varphi(\rho c_p)_s \quad (8)$$

where φ is the nanoparticle volume fraction, ρ_f is the density of the base fluid, ρ_s is the density of the solid particle, and c_p is the specific heat at constant pressure. The effective thermal conductivity of the nanofluid, corresponding to the Hamilton and Crosser model, is given by [28,30]:

$$\frac{k_{nf}}{k_f} = \frac{k_s + 2k_f - 2\varphi(k_f - k_s)}{k_s + 2k_f + \varphi(k_f - k_s)}, \quad (9)$$

where k_{nf}, k_f, and k_s are the thermal conductivities of the nanofluid, the fluid, and the solid particles, respectively.

Next, the non dimensional variables and functions are introduced as:

$$t = \frac{\nu_f}{L^2}t_1,\ y = \frac{y_1}{L},\ u = \frac{L}{\nu_f}u_1,\ \theta = \frac{T - T_\infty}{T_w - T_\infty},\ L = \left[\frac{\nu_f{}^2}{g\beta_f(T_w - T_\infty)}\right]^{1/3}, \quad (10)$$

Equations (2)–(6) take simplified dimensionless forms, as follows:

$$\frac{\partial u(y,t)}{\partial t} = \frac{1}{a_1}\frac{\partial^2 u(y,t)}{\partial y^2} + a_2\theta(y,t);\ y, t > 0 \quad (11)$$

$$\frac{\partial\theta(y,t)}{\partial t}=\frac{1}{a_3}\frac{\partial^2\theta(y,t)}{\partial y^2};\ y,t>0, \tag{12}$$

$$u(y,0)=0,\ \theta(y,0)=0;\ y\geq 0, \tag{13}$$

$$u(0,t)=0,\ \theta(0,t)=1;\ t>0, \tag{14}$$

$$u(y,t)\to 0,\ \theta(y,t)\to 0 \text{ as } y\to\infty, \tag{15}$$

where:

$$a_1=(1-\varphi)^{2.5}[(1-\varphi)+\varphi\tfrac{\rho_s}{\rho_f}],\ a_2=\frac{1-\varphi+\varphi\frac{(\rho\beta)_s}{(\rho\beta)_f}}{1-\varphi+\varphi\frac{\rho_s}{\rho_f}},$$
$$a_3=Pr\frac{1-\varphi+\varphi\frac{(\rho c_p)_s}{(\rho c_p)_f}}{\frac{k_{nf}}{k_f}+Nr},\ Pr=\frac{\mu_f c_{pf}}{k_f}, Nr=\frac{16\sigma^* T_\infty{}^3}{3k^* k_f}, \tag{16}$$

where Pr is the Prandtl number and Nr is the radiation parameter.

The fractional model of the nanofluid is described by the fractional differential equations:

$$^{c}D_t^{\alpha}u(y,t)=\frac{1}{a_1}\frac{\partial^2 u(y,t)}{\partial y^2}+a_2\theta(y,t);\ y,t>0, \tag{17}$$

$$^{c}D_t^{\alpha}\theta(y,t)=\frac{1}{a_3}\frac{\partial^2\theta(y,t)}{\partial y^2};\ y,t>0, \tag{18}$$

together with the initial and boundary conditions given by Equations (13)–(15). The operator $^{c}D_t^{\alpha}$ represents the Caputo time-fractional derivative, defined as [31]:

$$^{c}D_t^{\alpha}u(y,t)=\frac{1}{\Gamma(1-\alpha)}\int_0^t(t-s)^{-\alpha}\frac{\partial u(y,s)}{\partial s}ds;\ 0\leq\alpha<1. \tag{19}$$

The Caputo derivative is:

$$L\{^{c}D_t^{\alpha}u(y,t)\}(q)=q^{\alpha}\overline{u}(y,q)-q^{\alpha-1}u(y,0)\ if\ \overline{u}(y,q)=L\{u(y,t)\}(q) \tag{20}$$

and:

$$\lim_{\alpha\to 1}{}^{c}D_t^{\alpha}u(y,t)=\frac{\partial u(y,t)}{\partial t} \tag{21}$$

3. Solution of the Problem

In order to determine the solution of the fractional partial differential Equations (17) and (18), with the initial and boundary conditions (13)–(15), the Laplace transform technique will be used. Equation (18) is not coupled to the momentum equation. Consequently, we shall firstly determine the temperature field.

3.1. Determination of the Temperature Field

Applying the Laplace transform to Equation (18), and bearing in mind the corresponding initial and boundary conditions, we find that:

$$a_3 q^{\alpha}\overline{\theta}(y,q)=\frac{\partial^2\overline{\theta}(y,q)}{\partial y^2};\ y>0, \tag{22}$$

where q is the transform parameter and the Laplace transform $\overline{\theta}(y,q)$ of $\theta(y,t)$ has to satisfy the following conditions:

$$\overline{\theta}(0,q) = \frac{1}{q};\ \overline{\theta}(y,q) \to 0 \text{ as } y \to \infty. \tag{23}$$

The solution to the problems (22) and (23) is:

$$\overline{\theta}(y,q) = \frac{1}{q}\exp(-y\sqrt{a_3 q^{\alpha}});\ y > 0. \tag{24}$$

Applying the inverse Laplace transform to Equation (24), and using Equation (A1) from the Appendix A, we find that:

$$\theta(y,t) = \Psi\left(1, \frac{-\alpha}{2}; -y\sqrt{a_3}t^{-\frac{\alpha}{2}}\right) \text{ for } 0 < \alpha \leq 1, \tag{25}$$

where:

$$\Psi(a,-b;z) = \sum_{n=1}^{\infty} \frac{z^n}{n!\Gamma(a-nb)};\ b \in (0,1), \tag{26}$$

is the Wright function [32]. For $\alpha = 1$, Equation (25) becomes:

$$\theta(y,t) = erfc\left(\frac{y\sqrt{a_3}}{2\sqrt{t}}\right);\ y,t > 0. \tag{27}$$

Of course, a simple analysis clearly shows that this result is in accordance with that obtained by Turkyilmazoglu and Pop [13], Equation (3.19).

3.2. Calculation of the Velocity Field

Applying the Laplace transform to Equation (17), and taking into consideration the corresponding initial and boundary conditions, we find that:

$$\frac{\partial^2 \overline{u}(y,q)}{\partial y^2} - a_1 q^{\alpha}\overline{u}(y,q) = -a_1 a_2 \frac{1}{q}\exp(-y\sqrt{a_3 q^{\alpha}});\ y > 0 \tag{28}$$

where the Laplace transform $\overline{u}(y,q)$ of $u(y,t)$ has to satisfy the conditions:

$$\overline{u}(0,q) = 0;\ \overline{u}(y,q) \to 0 \text{ as } y \to \infty. \tag{29}$$

A particular solution of Equation (28) is:

$$\overline{u}_p(y,q) = \frac{a_1 a_2}{a_1 - a_3}\frac{1}{q^{\alpha+1}}\exp(-y\sqrt{a_3 q^{\alpha}}), \tag{30}$$

while its general solution is:

$$\overline{u}(y,q) = A\exp(y\sqrt{a_1 q^{\alpha}}) + B\exp(-y\sqrt{a_1 q^{\alpha}}) + \frac{a_1 a_2}{a_1 - a_3}\frac{1}{q^{\alpha+1}}\exp(-y\sqrt{a_3 q^{\alpha}}). \tag{31}$$

Considering the conditions of (29), it results that:

$$\overline{u}(y,q) = \tfrac{a_1 a_2}{a_1 - a_3}\tfrac{1}{q^{\alpha+1}}[\exp(-y\sqrt{a_3 q^{\alpha}}) - \exp(-y\sqrt{a_1 q^{\alpha}})];\ 0 < \alpha < 1. \tag{32}$$

Applying the inverse Laplace transform to Equation (32), using the convolution theorem and the Equality (A1), we find that:

$$u(y,t) = \frac{a_1a_2}{a_1-a_3}\frac{1}{\Gamma(1-\beta)}\int_0^t \frac{1}{(t-s)^{\beta}}\left[\Psi\left(1,\frac{\beta-1}{2};-y\sqrt{a_3s^{\beta-1}}\right)-\Psi\left(1,\frac{\beta-1}{2};-y\sqrt{a_1s^{\beta-1}}\right)\right]ds, \quad (33)$$

where $\beta = 1-\alpha$.

Now, using the identitiy (A2) from the Appendix A, we can present our solution in an interesting, but equivalent, form:

$$u(y,t) = \frac{a_1a_2}{a_1-a_3}\left\{{}^cD_t^{\beta}\left[t\Psi\left(2,\frac{\beta-1}{2};-y\sqrt{a_3t^{\beta-1}}\right)\right]-{}^cD_t^{\beta}\left[t\Psi\left(2,\frac{\beta-1}{2};-y\sqrt{a_1t^{\beta-1}}\right)\right]\right\}, \quad (34)$$

in terms of the Caputo derivative of the Wright functions.

3.3. Nusselt Number and Skin Friction

In order to determine the two entities of physical interest, namely the Nusselt number Nu and the skin friction coefficient C_f, we use the relations:

$$\begin{aligned} Nu = \frac{Lq_w}{k_f(T_w-T_\infty)} = -\frac{Lk_{nf}}{k_f(T_w-T_\infty)}\left.\frac{\partial T(y_1,t_1)}{\partial y_1}\right|_{y_1=0} = \\ -\frac{k_{nf}}{k_f}\left.\frac{\partial\theta(y,t)}{\partial y}\right|_{y=0} = -\frac{k_{nf}}{k_f}\lim_{y\to 0^+}L^{-1}\left\{\frac{\partial\bar{\theta}(y,q)}{\partial y}\right\}. \end{aligned} \quad (35)$$

$$\begin{aligned} C_f = \frac{\tau_w}{\rho_f\left(\frac{\nu_f}{t}\right)^2} = \frac{\mu_{nf}}{\rho_f\left(\frac{\nu_f}{t}\right)^2}\left.\frac{\partial u_1(y_1,t_1)}{\partial y_1}\right|_{y_1=0} = \\ \frac{\mu_{nf}}{\mu_f}\left.\frac{\partial u(y,t)}{\partial y}\right|_{y=0} = \frac{1}{(1-\varphi)^{2.5}}\lim_{y\to 0^+}L^{-1}\left\{\frac{\partial\bar{u}(y,q)}{\partial y}\right\}, \end{aligned} \quad (36)$$

where q_w is the constant heat flux from the surface of the plate and τ_w is the skin friction or shear stress on the boundary.

Introducing Equations (24) and (32) into (35) and (36), respectively, we find that:

$$Nu = \frac{k_{nf}}{k_f}\sqrt{a_3}\frac{t^{-\frac{\alpha}{2}}}{\Gamma(1-\frac{\alpha}{2})},\quad C_f = \frac{a_1a_2}{\sqrt{a_1}+\sqrt{a_2}}\frac{t^{\frac{\alpha}{2}}}{\Gamma(1+\frac{\alpha}{2})}\frac{1}{(1-\varphi)^{2.5}}. \quad (37)$$

Using the identity (A3) from the Appendix A, we also provide equivalent forms for Nu and C_f, namely:

$$Nu = 2\frac{k_{nf}}{k_f}\sqrt{\frac{a_3}{\pi}}\,{}^cD_t^{\frac{1+\alpha}{2}}(t^{1/2}),\quad C_f = \frac{1}{(1-\varphi)^{2.5}}\frac{a_1a_2}{\sqrt{a_1}+\sqrt{a_2}}\,{}^cD_t^{1-\frac{\alpha}{2}}(t), \quad (38)$$

in terms of the Caputo derivatives of $t^{1/2}$ and t.

4. Validation

In order to bring to light the accuracy of the results that have been obtained, it is suitable to show that they are in accordance with similar solutions from the existing literature. For that, let us use $\beta = 0$ in Equation (34). It corresponds to $\alpha = 1$, and the solution corresponding to the same unsteady natural convection flow of ordinary nanofluids, has to be obtained. When the Caputo derivative of zero order is the identity operator, Equation (34) becomes:

$$u(y,t) = \frac{a_1a_2}{a_1-a_3}\left\{\left[t\Psi\left(2,\frac{-1}{2};-y\sqrt{\frac{a_3}{t}}\right)\right]-\left[t\Psi\left(2,\frac{-1}{2};-y\sqrt{\frac{a_1}{t}}\right)\right]\right\}. \quad (39)$$

On the other hand (see also Equation (A2)):

$$erfc\left(\frac{y\sqrt{a}}{2\sqrt{t}}\right) = \Psi\left(1, \frac{-1}{2}; -y\sqrt{\frac{a}{t}}\right) = \frac{\partial}{\partial t}\left[t\Psi\left(2, \frac{-1}{2}; -y\sqrt{\frac{a}{t}}\right)\right] \tag{40}$$

and then:

$$t\Psi\left(2, \frac{-1}{2}; -y\sqrt{\frac{a}{t}}\right) = \int_0^t erfc\left(\frac{y\sqrt{a}}{2\sqrt{s}}\right)ds, \tag{41}$$

where $erfc(\cdot)$ is the complementary error fucntion of Gauss.

Now, by introducing Equation (41) in (39), and by using Equation (A4), we get the velocity field as the simple form:

$$u(y,t) = \frac{a_1 a_2}{a_1 - a_3}\left\{ \begin{array}{l} \left(t + \frac{y^2 a_3}{2}\right) erfc\left(\frac{y\sqrt{a_3}}{2\sqrt{t}}\right) - \left(t + \frac{y^2 a_1}{2}\right) erfc\left(\frac{y\sqrt{a_1}}{2\sqrt{t}}\right) + \\ \frac{y\sqrt{a_1 t}}{\sqrt{\pi}} \exp\left(-\frac{y^2 a_1}{4t}\right) - \frac{y\sqrt{a_3 t}}{\sqrt{\pi}} \exp\left(-\frac{y^2 a_3}{4t}\right) \end{array} \right\}. \tag{42}$$

Finally, bearing in mind the notations of Turkyilmazoglu and Pop [13], as well as their rescaling relation from equality (2.11), it is easy to show that our solution (42) is identical to Equation (3.20), from [13].

With regards to the Nusselt number Nu and the skin friction coefficient C_f, we use $\alpha = 1$ in Equation (38) and use Equation (A5). The expressions corresponding to the ordinary nanofluid are:

$$Nu = \frac{k_{nf}}{k_f}\frac{\sqrt{a_3}}{\sqrt{\pi t}},\ C_f = \frac{1}{(1-\varphi)^{2.5}}\frac{2a_1 a_2}{\sqrt{a_1}+\sqrt{a_3}}\sqrt{\frac{t}{\pi}} \tag{43}$$

As expected, by changing $a_1, a_3,$ and t, by $\frac{1}{a_1}, \frac{1}{a_3},$ and τ_1, respectively, we recover the solutions (3.21), from [13].

5. Numerical Results and Discussion

The natural convection flow of water-based fractional nanofluids over an infinite vertical plate with thermal radiation and a uniform temperature on the boundary, is analytically studied. Closed form solutions for the dimensionless temperature and velocity fields, and the two entities of physical interest, the Nusselt number and skin friction coefficient, are determined in equivalent forms, in terms of the Wright function or its fractional derivatives. It is worth pointing out that all of these solutions have been immediately reduced to the known solutions, based on the literature for ordinary nanofluids. A table containing the thermo-physical properties of copper oxide (CuO) and silver (Ag) is also included for later use.

In order to bring to light the influence of the fractional parameter on the heat and mass transfer in the natural convection flow of the above-mentioned fractional nanofluids, and therefore to obtain some physical insight into the present results, some numerical calculations have been carried out for different values of the fractional parameter α, radiation parameter Nr, and the nanoparticle value fraction φ. For comparison, the diagrams of dimensionless temperature and velocity fields, and the Nusselt number and skin friction coefficient corresponding to fractional nanofluids (for different values of the fractional parameter $\alpha \in (0,1)$) and those of ordinary nanofluids (when $\alpha = 1$), are depicted in Figures 1–4. As was expected, in all cases, the diagrams corresponding to the fractional nanofluids tend to superpose over those of ordinary nanofluids, when $\alpha \to 1$.

Profiles of the dimensionless temperature $\theta(y,t)$ against y are presented in Figure 1a–c, for the different values of the fractional parameter α and the CuO nanoparticle volume fraction φ. The fluid temperature, as it results from these figures, increases with respect to α, up to a critical value of y (less than 0.5), and then decreases. Consequently, in terms of the plate proximity, the heat transfer is stronger when the thermal boundary layer is thinner for fractional nanofluids, in comparison to the ordinary ones. An opposite trend appears at a later point in time. With respect to the volume fraction φ, the temperature is an increasing function for both ordinary and fractional nanofluids, and it smoothly decreases from the maximum value of one on the boundary, to the zero value far away from the plate. In Table 2, the temperature at different values of y and of the fractional parameter α, is given for nanofluids containing nanoparticles of CuO or Ag, and these values are in full accordance with those resulting from Figure 1. Furthermore, as results from this table, the Ag nanoparticles induce larger temperature values and a lower heat transfer to ordinary or fractional nanofluids, in comparison to CuO nanoparticles.

The effect of enhancing the heat transfer rate against φ with the fractional parameter α, for CuO-water and Ag-water nanofluids, is presented in Figure 2, in the absence or presence of thermal radiation. For both fractional nanofluids, the Nusselt number Nu is an increasing function with respect to α. It is also an almost linearly increasing function of φ. Further, the heat transfer rate for the CuO-water fractional nanofluid is always a little higher than that corresponding to Ag-water fractional nanofluid. However, the difference between them increases with increasing values of φ.

The influence of the fractional parameter α and of the nanoparticle volume fraction φ, on the dimensionless velocity $u(y,t)$ against y, is brought to light by Figure 3a–c; which is also seen for a nanofluid with CuO nanoparticles. Near the plate, the nanofluid velocity increases up to a maximum value and then asymptotically decreases to the zero value for y values greater than 2.5, but it is a decreasing function with respect to the two parameters α and φ. Consequently, the boundary layer thickness is lower for ordinary nanofields, in comparison to fractional nanofluids. From a physical point of view, it means that the nanofluid viscosity decreases for increasing values of α or φ, and the ordinary nanofluids exhibit a stronger capacity in flow. This implies that the viscoelasticity strengthens the flow resistance with a decrease in the fractional parameter [33]. The variation of the skin friction coefficient C_f, against the nanoparticle volume fraction φ, is presented in Figure 4, for the same nanofluids at different values of α and Nr, equal to zero or one.

Table 2. Values of the dimensionless temperature for $\varphi = 0.2$, $t = 0.5$, and different values of the fractional parameters α and y.

y	CuO				Ag			
	$\alpha = 0.1$	$\alpha = 0.5$	$\alpha = 0.85$	$\alpha = 1$	$\alpha = 0.1$	$\alpha = 0.5$	$\alpha = 0.85$	$\alpha = 1$
0	1	1	1	1	1	1	1	1
0.1	0.86341	0.84348	0.84816	0.87689	0.86675	0.84694	0.85013	0.87864
0.2	0.74363	0.72536	0.75925	0.77086	0.74937	0.73083	0.76448	0.7768
0.3	0.64057	0.62803	0.65475	0.6609	0.648	0.63504	0.66273	0.6691
0.4	0.55187	0.54341	0.55664	0.55868	0.56041	0.55166	0.56628	0.56886
0.5	0.47551	0.46851	0.46887	0.46476	0.48472	0.47766	0.47961	0.47635
0.6	0.40976	0.40218	0.39111	0.38035	0.4193	0.4119	0.4025	0.39275
0.7	0.35313	0.34379	0.32302	0.30608	0.36274	0.35376	0.33461	0.31872
0.8	0.30436	0.29277	0.26415	0.24213	0.31383	0.30272	0.27556	0.25448
0.9	0.26234	0.2485	0.21388	0.18821	0.27154	0.25821	0.22481	0.19985
1	0.22613	0.2103	0.17148	0.14371	0.23496	0.2196	0.1817	0.15432

The variation of the skin friction coefficient C_f, against the nanoparticle volume fraction φ, is presented in Figure 4, for the same nanofluids at different values of α and Nr, equal to zero or one. In both cases, i.e., in the absence or presence of radiation, the skin friction coefficient is a decreasing function with respect to the fractional parameter α. Considering its variation with respect to φ, two different situations appear. In the absence of radiation (Figure 4b), it is an increasing function with respect to φ, and its values are always greater for Ag-water fractional nanofluid compared to CuO-water fractional nanofluid. In the presence of radiation, when $Nr = 1$ (Figure 4a), the shear stress on the plate decreases up to a critical value of φ (about 0.1), and then increases. Furthermore, up to this value of φ, it is smaller for Ag-water fractional nanofluid in comparison to CuO-water fractional nanofluid, and a reverse situation is presented at a later point in time. Throughout this study, the value of Pr has been taken as 6.067.

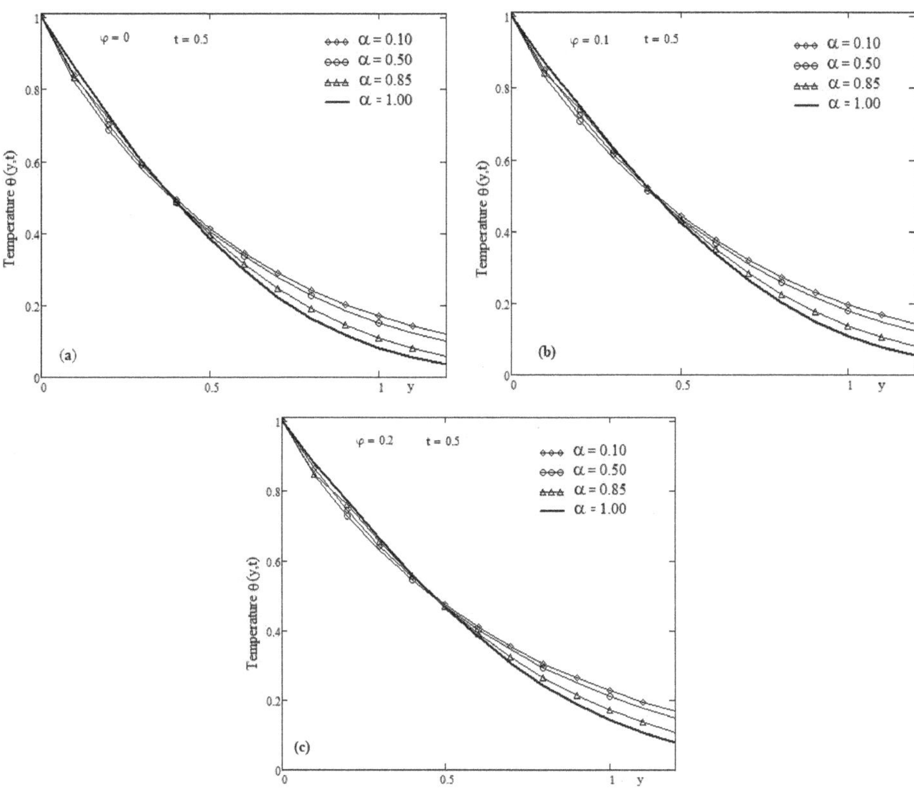

Figure 1. Variation of the dimensionless temperature $\theta(y,t)$ with the fractional parameter α and the CuO nanoparticle volume fraction φ. (**a**) for $\varphi = 0$; (**b**) for $\varphi = 0.1$; (**c**) for $\varphi = 0.2$.

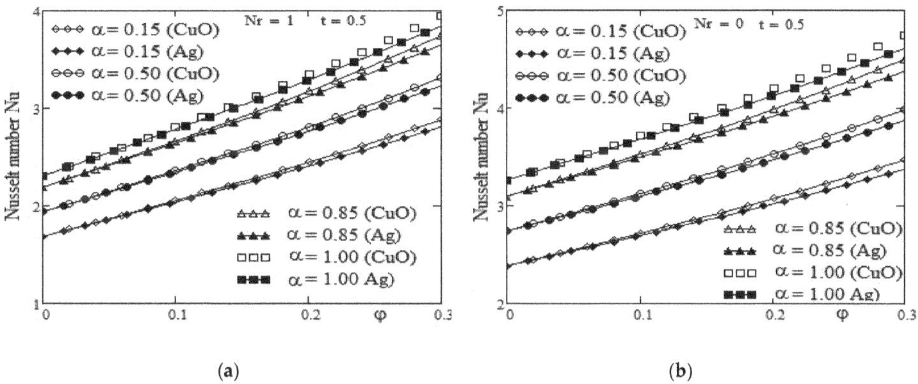

Figure 2. Values of the Nusselt number against the nanoparticle volume fraction φ for the two water-based nanofluids. (**a**) for $Nr = 1$; (**b**) for $Nr = 0$.

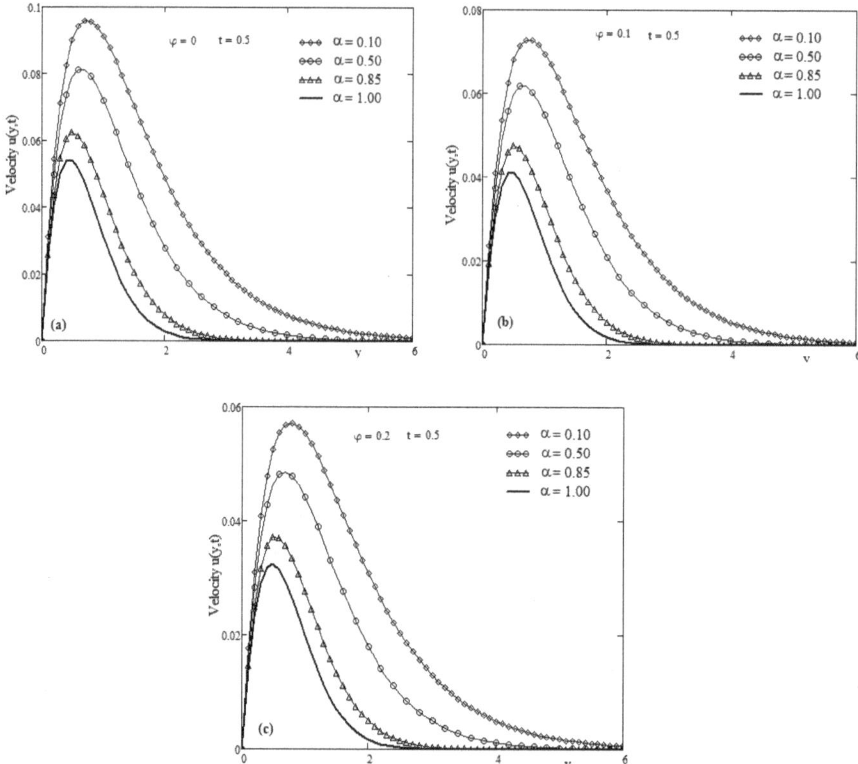

Figure 3. Variation of the velocity $u(y,t)$ with the fractional parameter α and the nanoparticle volume fraction φ. (**a**) for $\varphi = 0$; (**b**) $\varphi = 0.1$; (**c**) for $\varphi = 0.2$.

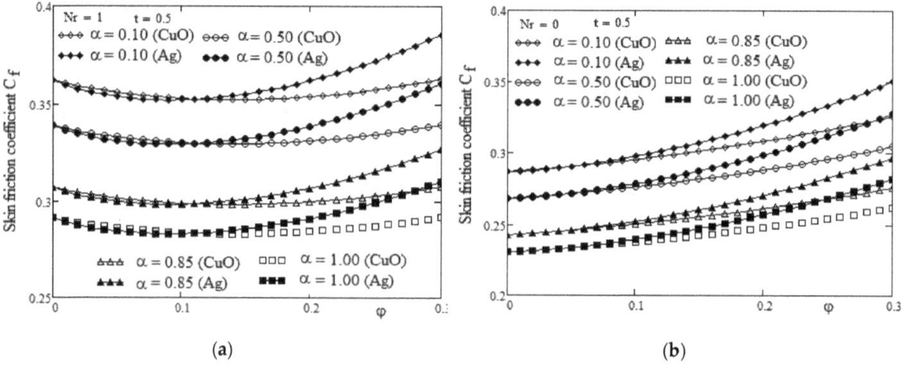

Figure 4. Values of the skin friction coefficient C_f against the nanoparticle volume fraction φ for two water-based fractional nanofluids. (**a**) for $Nr = 1$; (**b**) for $Nr = 0$.

6. Conclusions

An analytical study of the natural convection flow of some water-based fractional nanofluids over an infinite vertical plate with thermal radiation and a uniform temperature on the boundary, is developed using the Caputo time-fractional derivative. The closed forms of solutions for the dimensionless temperature and velocity fields, and the corresponding Nusselt number and skin friction coefficient, are established in equivalent forms, in terms of the Wright function and its fractional derivatives. They have been reduced to the solutions obtained in [13], corresponding to ordinary nanofluids, when the fractional parameter $\alpha \to 1$.

In order to get some physical insight into the results, which have been obtained for CuO-water and Ag-water fractional nanofluids, some numerical calculations and graphical representations have been presented in Table 2 and Figures 1–4, for different values of the fractional parameter α, radiative parameter Nr, and the nanoparticles volume fraction φ. The main findings are:

The enhancement of heat transfer in the natural convection flows is lower for fractional nanofluids, in comparison to ordinary nanofluids. The thermal boundary layer is thicker for fractional nanofluids. In both cases, the fluid temperature increases and the heat transfer declines for increasing values of the nanoparticle volume fraction φ.

The flows of water-based fractional nanofluids are faster than the ordinary nanofluids.

A decrease in the fractional derivative parameter increases the thickness of the velocity boundary layer. From a physical point of view, it means that the nanofluid viscosity increases with the decreasing of α.

The dimensionless velocity of water-based fractional nanofluids, as well as that of ordinary nanofluids, is a decreasing function with respect to the nanoparticle volume fraction φ.

The skin friction coefficient in the natural convection flow is higher for fractional nanofluids, in comparison to ordinary nanofluids.

The enhancement of heat transfer in natural convection is stronger for Cu-water fractional nanofluids, when compared with Ag-water fractional nanofluids.

In the presence of radiation, Ag-water fractional/ordinary nanofluids achieve a lower skin friction coefficient near the plate, relative to Cu-water fractional/ordinary nanofluids.

Our study shows that the nanofluids described by fractional derivatives have a significantly different behavior than ordinary nanofluids. The same conclusions were obtained in the references [33,34]. Even if the models studied in these articles are different from the model studied here, the obtained results showed that the fractional parameter has a strong influence on the heat transfer process.

Acknowledgments: Waqas Ali Azhar is grateful to Abdus Salam School of Mathematical Sciences, Government College University, Lahore, Pakistan for supporting this scientific research.

Conflicts of Interest: The authors declare no conflict of interest.

Appendix A

$$L^{-1}\left\{\frac{1}{q^a}\exp(-cq^b)\right\} = t^{a-1}\Psi\left(a,-b;-ct^{-b}\right);\ b\in(0,1) \tag{A1}$$

$$\frac{\partial}{\partial t}\left[t\Psi\left(2,-b;-ct^{-b}\right)\right] = \Psi(1,-b;-ct^{-b});\ b\in(0,1) \tag{A2}$$

$$^{c}D_t^{\alpha}(t^{\gamma}) = \frac{\Gamma(\gamma+1)}{\Gamma(\gamma+1-\alpha)}t^{\gamma-\alpha} \tag{A3}$$

$$\int_0^t erfc\left(\frac{y\sqrt{a}}{2\sqrt{s}}\right)ds = \left(t+\frac{ay^2}{2}\right)erfc\left(\frac{y\sqrt{a}}{2t}\right) - \frac{y\sqrt{a}}{\sqrt{\pi}}\exp\left(\frac{-ay^2}{4t}\right) \tag{A4}$$

$$^{c}D_t^{1}(t^{1/2}) = \frac{1}{2\sqrt{t}},\ ^{c}D_t^{1/2}(t) = \frac{2}{\sqrt{\pi}}\sqrt{t} \tag{A5}$$

References

1. Raptis, A. Unsteady free convection flow through a porous medium. *Int. J. Eng. Sci.* **1983**, *21*, 345–348. [CrossRef]
2. Kim, S.J.; Vafai, K. Analysis of natural convection about a vertical plate embedded in a porous medium. *Int. J. Heat Mass Transfer* **1989**, *32*, 665–677. [CrossRef]
3. Narasimha, K.R.; Pop, I. Transient free convection in a fluid saturated porous media with temperature dependent viscosity. *Int. Commun. Heat Mass Transf.* **1994**, *21*, 573–581.
4. Ghosh, S.K.; Beg, O.A. Theoretical analysis of radiative effects on transient free convection heat transfer past a hot vertical surface in porous media. *Nonlinear Anal. Model. Control* **2008**, *13*, 419–432.
5. Mondal, S.; Haroun, N.A.H.; Sibanda, P. The effects of thermal radiation on an unsteady MHD axisymmetric stagnation-point flow over a shrinking sheet in presence of temperature dependent thermal conductivity with Navier slip. *PLoS ONE* **2015**, *10*, e0138355. [CrossRef] [PubMed]
6. Zhang, C.; Zheng, L.; Zhang, X.; Chen, G. MHD flow and radiation heat transfer of nanofluids in porous media with variable surface heat flux and chemical reaction. *Appl. Math. Model.* **2015**, *39*, 165–181. [CrossRef]
7. Abd Elazem, N.Y.; Ebaid, A.; Aly, H.E. Radiation effect of MHD on Cu-water and Ag-water nanofluids flow over a stretching sheet: Numerical study. *J. Appl. Computat. Math.* **2015**, *4*. [CrossRef]
8. Bianco, V.; Manca, O.; Nardini, S.; Vafai, K. *Heat Transfer Enhancement with Nanofluids*; CRC Press: London, UK, 2015.
9. Choi, S.U.S. Enhancing thermal conductivity of fluids with nanoparticles. In Proceedings of the ASME International Mechanical Engineering Congress and Exposition 66, San Francisco, CA, USA, 12–17 November 1995; pp. 99–105.
10. Khan, W.A.; Aziz, A. Natural convection flow of a nanofluid over a vertical plate with uniform surface heat flux. *Int. J. Therm. Sci.* **2011**, *50*, 1207–1214. [CrossRef]
11. Turkyilmazoglu, M. Exact analytical solutions for heat and mass transfer of MHD slip flow in nanofluids. *Chem. Eng. Sci.* **2012**, *84*, 182–187. [CrossRef]
12. Bachok, N.; Ishak, A.; Pop, I. Flow and heat transfer characteristics on a moving plate in a nanofluid. *Int. J. Heat Mass Transf.* **2012**, *55*, 642–648. [CrossRef]
13. Turkyilmazoglu, M.; Pop, I. Heat and mass transfer of unsteady natural convection flow of some nanofluids past a vertical infinite flat plate with radiation effect. *Int. J. Heat Mass Transf.* **2013**, *59*, 167–171. [CrossRef]
14. Mohankrishna, P.; Sugunamma, V.; Sandeep, N. Radiation and magnetic field effects on unsteady natural convection flow of a nanofluid past an infinite vertical plate with heat source. *Chem. Process. Eng. Res.* **2014**, *25*, 39–52.
15. Ellahi, R. The effects of MHD and temperature dependent viscosity on the flow of non-Newtonian nanofluid in a pipe: Analytical solutions. *Appl. Math. Model.* **2013**, *37*, 1451–1457. [CrossRef]
16. Rizwan, U.H.; Khan, Z.H.; Hussain, S.T.; Hammouch, Z. Flow and heat transfer analysis of water and ethylene glycol based Cu nanoparticles between two parallel disks with suction/injection effects. *J. Mol. Liq.* **2016**, *221*, 298–304. [CrossRef]
17. Sheikholeslami, M.; Zaigham, Q.M.; Ellahi, R. Influence of induced magnetic field on free convection of nanofluid considering Koo-Kleinstreuer-Li (KKL) correlation. *Appl. Sci.* **2016**, *324*, 1–13. [CrossRef]
18. Sekrani, G.; Poncet, S. Further investigation on laminar forced convection of nanofluid flows in a uniformly heated pipe using direct numerical simulations. *Appl. Sci.* **2016**, *332*, 1–24. [CrossRef]
19. Rizwan, U.H.; Prabhakar, B.; Bandari, S. Impact of inclined Lorentz forces on Tangent hyperbolic nanofluid flow with zero normal flux of nanoparticles at the stretching sheet. *Neural Comput. Appl.* **2016**. [CrossRef]
20. Ibrahim, W.; Ul Haq, R. Magnetohydrodynamic (MHD) stagnation point flow of nanofluid past a stretching sheet with convective boundary condition. *J. Braz. Soc. Mech. Sci. Eng.* **2016**, *38*, 1155–1164. [CrossRef]
21. Waqar, A.; Khan, C.R.; Rizwan, U.H. Heat Transfer Analysis of MHD Water Functionalized Carbon Nanotube Flow over a Static/Moving Wedge. *J. Nanomater.* **2015**, *16*, 934367. [CrossRef]
22. Khan, W.A.; Khan, Z.H.; Haq, R.U. Flow and heat transfer of ferrofluids over a flat plate with uniform heat flux. *Eur. Phys. J. Plus* **2015**, *130*. [CrossRef]
23. Rizwan, U.H.; Rajotia, D.; Noor, N.F.M. Thermophysical effects of water driven copper nanoparticles on MHD axisymmetric permeable shrinking sheet: Dual-nature study. *Eur. Phys. J.* **2016**, *39*. [CrossRef]

24. Sheikholeslami, M.; Ellahi, R. Electrohydrodynamic nanofluid hydrothermal treatment in an enclosure with sinusoidal upper wall. *Appl. Sci.* **2015**, *5*, 294–306. [CrossRef]
25. Rahman, S.U.; Ellahi, R.; Nadeem, S.; Zia, Q.M. Simultaneous effects of nanoparticles and slip on Jeffrey fluid through tapered artery with mild stenosis. *J. Mol. Liq.* **2016**, *218*, 484–493. [CrossRef]
26. Sheikholeslami, M.; Ellahi, R. Three dimensional mesoscopic simulation of magnetic field effect on natural convection of nanofluid. *Int. J. Heat Mass Transf.* **2015**, *89*, 799–808. [CrossRef]
27. Akbar, N.S.; Raza, M.; Ellahi, R. Copper oxide nanoparticles analysis with water as base fluid for peristaltic flow in permeable tube with heat transfer. *Comput. Meth. Prog. Biomed.* **2016**, *130*, 22–30. [CrossRef] [PubMed]
28. Oztop, H.F.; Abu-Nada, E. Numerical study of natural convection in partially heated rectangular enclosures filled with nanofluids. *Int. J. Heat Fluid Flow* **2008**, *29*, 1326–1336. [CrossRef]
29. Tiwari, R.K.; Das, M.K. Heat transfer argument in a two-sided lid-driven differentially heated square cavity utilizing nanofluids. *Int. J. Heat Mass Transf.* **2007**, *50*, 2002–2018. [CrossRef]
30. Kakac, S.; Pramuanjaroenkij, A. Review of convective heat transfer enhancement with nanofluids. *Int. J. Heat Mass Transf.* **2009**, *52*, 3187–3196. [CrossRef]
31. Povstenko, Y. *Linear Fractional Diffusion-Wave Equation for Scientists and Engineers*; Springer: Basel, Switzerland, 2015.
32. Stankovic, B. On the function of E. M. Wright. *Publ. I'Inst. Math.* **1970**, *10*, 113–124.
33. Cao, Z.; Zhao, J.; Wang, Z.; Liu, F.; Zheng, L. MHD flow and heat transfer of fractional Maxwell viscoelastic nanofluid over a moving plate. *J. Mol. Liq.* **2016**, *222*, 1121–1127. [CrossRef]
34. Pan, M.; Zheng, L.; Liu, F.; Zhang, X. Modeling heat transport in nanofluids with stagnation point flow using fractional calculus. *Appl. Math. Model.* **2016**, *40*, 8974–8984. [CrossRef]

Article

Magnetohydrodynamic Nanoliquid Thin Film Sprayed on a Stretching Cylinder with Heat Transfer

Noor Saeed Khan [1], Taza Gul [1,*], Saeed Islam [1], Ilyas Khan [2], Aisha M. Alqahtani [3] and Ali Saleh Alshomrani [4]

1. Department of Mathematics, Abdul Wali Khan University, Mardan 32300, Khyber Pakhtunkhwa, Pakistan; noorsaeedkhankhattak@gmail.com (N.S.K.); saeedislam@awkum.edu.pk (S.I.)
2. Basic Engineering Sciences Department, College of Engineering, Majmaah University, Majmaah 11952, Saudi Arabia; ilyaskhanqau@yahoo.com
3. Department of Mathematics, Princess Nourah bint Abdulrahman University, Riyadh 11564, Saudi Arabia; Alqahtani@pnu.edu.sa
4. Department of Mathematics, Faculty of Science, King Abdulaziz University, Jeddah, 21577, Saudi Arabia; aszalshomrani@kau.edu.sa

* Correspondence: tazagulsafi@yahoo.com; Tel.: +92-331-911-7160

Academic Editor: Rahmat Ellahi
Received: 31 January 2017; Accepted: 3 March 2017; Published: 10 March 2017

Abstract: The magnetohydrodynamic thin film nanofluid sprayed on a stretching cylinder with heat transfer is explored. The spray rate is a function of film size. Constant reference temperature is used for the motion past an expanding cylinder. The sundry behavior of the magnetic nano liquid thin film is carefully noticed which results in to bring changes in the flow pattern and heat transfer. Water-based nanofluids like Al_2O_3-H_2O and CuO-H_2O are investigated under the consideration of thin film. The basic constitutive equations for the motion and transfer of heat of the nanofluid with the boundary conditions have been converted to nonlinear coupled differential equations with physical conditions by employing appropriate similarity transformations. The modeled equations have been computed by using HAM (Homotopy Analysis Method) and lead to detailed expressions for the velocity profile and temperature distribution. The pressure distribution and spray rate are also calculated. The comparison of HAM solution predicts the close agreement with the numerical method solution. The residual errors show the authentication of the present work. The CuO-H_2O nanofluid results from this study are compared with the experimental results reported in the literature showing high accuracy especially, in investigating skin friction coefficient and Nusselt number. The present work discusses the salient features of all the indispensable parameters of spray rate, velocity profile, temperature and pressure distributions which have been displayed graphically and illustrated.

Keywords: magnetohydrodynamic; nanoliquid thin film; Al_2O_3-H_2O; CuO-H_2O; spray; heat transfer; stretching cylinder; homotopy analysis method; numerical method; residual errors; skin friction coefficient; nusselt number

1. Introduction

Recently heat transfer technology is related with the cooling applications of miniaturized high heat flux components. The fluids traditionally used for heat transfer applications such as water, oils and ethylene glycol have comparatively low thermal conductivity and do not meet the required demand as an efficient heat transfer agent. The conventional technique for increasing heat dissipation is to increase the area available for exchanging heat with a heat transfer fluids, but this process needs an undesirable increase in the size of thermal management system. So by taking into account the rising demands of modern technology, comprising chemical production, power stations and microelectronics, there is a need to propose new types of fluids which are more effective with respect to heat exchange performance. For this purpose nanofluids are constructed to ensure effective thermal conductivity enhancements and to fulfill the rising demands of cooling/heating and other needs. Nanofluid is a dispersion consisting of nanometer-sized particles, called naoparticles. Nanoliquids are dispersions formulated as a violent encounter of moving nanoparticles in a base liquid. These particles utilized in nanofuids are generally constructed from metallic elements (Al, Cu), oxides (Al_2O_3), nitrides (AlN, SiN) or nonmetallic elements (graphite, carbon nanotubes) and a liquid of conduction nature like water or ethylene glycol usually used as the base fluid. Oily products, biofluids and polymer suspensions may are used as base fluid. Nanoparticles have the diameters in the range 1 and 100 nm. For ensuring improved transfer of heat enhancements, nanofluids generally include up to 5 % volume fraction of nanoparticles. Due to strange characteristics of nanoliquids that make them capable in many applications in hybrid-powered engines, pharmaceutical processes, fuel cells, including microelectronics and heat transfer. It has been widely used in engineering devices for ship-sand in boiler flue gas reduction of temperature and defense, in space, in machining, in grinding, nuclear reactor and heat exchanger, chiller, domestic refrigerator and engine vehicle thermal/cooling management. Nanofluids makes better the thermal conduction of the base liquid extremely, therefore in the analysis of flow of nanofluids the researchers are deeply interested. Nanofluids are also very consistent and have no supplementary issues like additional pressure drop, erosion and sedimentation. For the first time Choi [1] introduced the nanofluid technology. Utilizations in the superconducting magnets and super fast computing are facing problems in thermal management. Thermal properties of nanofluids are extensive research area during the past few decades for their perspective technological applications in electronics cooling and heat transfer. Yu et al. [2] analyzed the thermal transport properties of ethylene glycol-based nanofluids experimentally by measuring the thermal conductivity and viscosity. In another study Yu et al. [3] investigated experimentally the rheological effects and heat transfer properties of Al_2O_3 nanofluids based on the mixtures 45 vol. % ethylene glycol and 55 vol. % water. With the discovery of carbon nanotubes (CNTs) in 1991, carbon-based nanomaterials are popular for their unique physical, thermal, mechanical and electrical properties. Chen et al. [4] employed a green method to make composites of multi-walled carbon nanotubes (MWNTs) decorated with silver nanoparticles (Ag-NPs). In case of graphene, a single atomic layer of graphite, which is a two dimensional form of carbon, is found to show good crystal quality and to have ballistic electronic transport at room temperature. It has been evolving as a fascinating material having unique physical, chemical and mechanical properties as investigated by Xie and Chen [5]. Nanofluids containing graphene oxide nanosheets have substantially higher thermal conductivities than the base fluid as analyzed by Yu et al. [6]. It is very necessary to know more about heat transfer properties of fluids due to the different speculated applications. It is a fact that the study of heat transfer techniques can enhance the comprehensive understanding of physical phenomena like convection and boiling. The effective thermal conductivity is one of the most controversial topics in nanofluids. In addition, the physical nature is still far from being well understood due to its complexity and diversity. After detailed analysis, Brownian motion induced convection and effective conduction through percolating nanoparticle paths are taken to be the two most probable mechanisms that yield the improved heat conduction in nanofluids. Through a mechanistic point of view, although the effects of some parameters such as the average size of nanoparticles and nanoparticles concentration have been discussed in literatures [1–4], yet an overall mechanistic description is not available. It is a problem to investigate the effective thermal conductivity of nanofluids analytically due to the extremely complicated mechanisms of heat transfer and the interrelationship between thermal conductivity and the size of nanoparticles and the nanoparticles concentration. Therefore the clear mechanism of heat transfer is still not known in nanofluids. To fill this gap Xiao et al. [7] attempted to derive the analytical expressions for effective thermal conductivity of nanofluids while taking into account the effect of heat convection due to the Brownian motion of nanoparticles based on the fractal theory. Similarly Cai et al. [8] investigated the advances of nanoparticles researches by introducing the fractal theory presenting the fractal model of thermal conductivity of nanofluids under the consideration of fractal distribution of nanoparticles sizes and heat convection between nanoparticles and liquids due to the Brownian motion of nanoparticles in fluids, in which the nanoparticles are assumed to be dispersed. Buongiorno [9] considered a detailed discussion about the nanofluid. Ellahi [10] explored the effects of MHD and temperature dependent viscosity on the flow of a non-Newtonian nanofluid in a pipe. Khan and Pop [11] elaborated the motion of a nanofluid passing a stretching surface. The effects of wall behavior for the peristaltic flow of a nanofluid were described by Mustafa et al. [12]. Akbar and Nadeem [13] investigated the endoscopic exploration of peristaltic flow of a nanofluid. Nowar [14] solved the problem of peristaltic flow of a nanofluid in the regime of Hall current and porous medium. A vast study exploring different aspects of nanofluid may be consulted in the references [15–23].

A magnetic nanofluid is a special substance having the combined features of fluid and magnetic characteristics. These fluids are practiced resulting in numerous distinguished utilizations like magneto-opticle wavelength filters, and other substances related to optic like nonlinear materials and tunable fiber filters, gratings and switches. Several types of physical characteristics of such fluids are changed by the help of varying the intensity of magnetic field. Nanofluids related to magnetism assume presently a broad application in several departments such as biomedicine, medicine and in sink float isolation. Too many biomedical uses that contains nanofluids, like drug supply, magnetic cell separation and negative growth in magnetic resonance imaging are most important. Magnetohydrodynamic (MHD) motions are significant due to their use in power generators, MHD accelerators, refrigerations coils, transmissions lines, electric transformers and heating elements. Due to the scope of this notion, several researchers have done their work on magnetohydrodynamic motions. It is a proved statement that energy transfer is possible due to composition (concentration) gradient and mass transportation is occurred due to temperature gradient. These contributions of MHD motion are utilized in levitation and pumping of fluids in mechanical engineering, controlling of liquid motion and transpiration procedures and aerodynamics. In extrusion process, investigation of transfer of heat in boundary layer flow past the stretching surfaces has wide uses. The cooling technique has to be controlled completely because the standard of final product relies the rate of transfer of heat. The desired quality of manufactured product may be achieved by the magnetohydrodynamic motion in electrically conducting liquid which can control the influence of cooling. Important industrial applications of the problem of viscous motion and transfer of heat past a stretching sheet are, for example, in few technical procedures like glass fiber production, manufacturing of foods and paper, continuous casting, wire drawing, in metallurgical techniques, such as crystal making, preparing of plastic and rubber sheets, enameling and painting of bronze threads and many others. The issues in molten form after some time stretched from a slit to obtain the required size during the manufacturing of these sheets. Due to the cooling stretching rate in the process and the process of stretching the final product with the required properties is prepared. Chamkha et al. [24] analyzed the Brownian motion and thermophoretic influences on the mixed convection MHD motion of a nanoliquid past a stretching porous medium. During suction/injection the discussion of time dependent magnetohydrodynamic motion of a nanoliquid past a vertical stretching surface is carried out by Kandasmy et al. [25]. Enough research work on motion past stretching surfaces has been carried out. Sakiadis [26] was the first one who considered the flow on continuous flat and solid surfaces. Crane [27] investigated the flow past an extensible surface by assuming the surface velocity varying in a linear way with an extent of space from the slit. Vajravelu et al. [28] determined the solution of the problem of magnetohydrodynamic motion of a non-Newtonian liquid past a stretching surface. Abu-Nada [29] examined the impacts of variable properties of Al_2O_3 water nanoliquid on the improvement of transfer of heat in natural convection. Nasrin and Alim [30] investigated the heat generation by nanofluid of variable properties with flat plate solar collector.

Thin film fluid flows are the subject of considerable attention in research. Khan et al. [31] reported the influences of variable properties of a thin liquid film motion past a contracting/expanding space. Recently, The behavior of flow and transfer of heat of a second-grade fluid through a porous medium past a stretching sheet is discussed by Khan et al. [32]. In another study, Khan et al. [33] investigated thermophoresis and thermal radiation with heat and mass transfer in a magnetohydrodynamic thin film second grade fluid of variable properties past a stretching sheet. Aziz et al. [34] reported thin film motion and transfer of heat on a time dependent stretching sheet with internal heating. The thin film Williamson nanofluid flow with varying viscosity and thermal conductivity on a time dependent sheet was analyzed by Khan et al. [35]. Qasim et al. [36] discussed transfer of heat and mass in a nanofluid past a time dependent stretching surface using Buongiorno's model. Prashant et al. [37] presented transfer of heat on a time dependent stretching surface with thermal radiation, internal heating in case of external magnetic field for a thin film flow. Kumari et al. [38] investigated transfer of heat and mass in a nanofluid past an unsteady stretching surface.

During past several years, a prominent revolution has been occurred in the analysis of motion and transfer of heat past slender cylinders. The inevitable needs of the utilization of slender objects which minimizes the drag and producing complete lift to support the body in some situations. In case of slender cylinder, the radius is possible to have same as the boundary layer thickness and the motion is taken as axisymmetric instead of two dimensional and the transverse curvature term existing in the governing equation which by force affects the velocity and temperature fields. Related to this notion, the coefficient of skin friction and rate of transfer of heat at the wall are affected too by the normal curvature. Motion past a cylinder and the corresponding transfer of heat characteristics have extensive applications like preparing of heat exchanger tubes, chimney stacks, cooling towers, offshore structure, thin film condensation and paper production. It is also utilized in petroleum industries, plasma studies and geothermal energy extractions. Wang [39] was the the first to analyze the flow in viscous case past a stretching hollow cylinder by evaluating the third order non-linear complicated system. The motion of a fluid outside a stretching cylinder by using Keller-box technique for the solution is elaborated by Ishak et al. [40]. Wang [41] obtained the similarity solution of a natural convective study over a non-horizontal stretching cylinder. The numerically solution of the problem of MHD motion and transfer of heat of Newtonian liquid past a stretching cylinder is obtained by Elbashbeshy [42]. Ashorynejad et al. [43] reported the motion of a nanofluid with transfer of heat and magnetic field past a stretching cylinder. Rangi and Ahmad [44] analyzed the motion and transportation of heat of a Newtonian incompressible liquid past a stretching cylinder with variable fluid property. Sheikholeslami [45] discussed the effect of uniform suction of a nanofliquid motion and transfer of heat past a stretching cylinder. Wang [46] explored the liquid film sprayed on a stretching cylinder. Koo and Kleinstreuer [47] described the viscous dissipation effects in micro channels. Koo [48] investigated the computational fluid flow with transfer of heat used in micro systems. Other investigations about nanofluids and different interesting problems with respect to different aspects are exist in the references [49–54].

The literature survey shows that there is a large amount of investigations about the nanofluids. Still there is no exploration to discuss the magnetohydrodynamic thin liquid film sprayed on a stretching cylinder with transfer of heat features of CuO-H_2O nanofluid. To fill this gape in this regard, claim is exist for the first try to analyze the sprayed liquid thin film effect on magnetohydrodynamic two dimensional CuO-H_2O nanofluid past a stretching cylinder with heat transfer. Employing appropriate transformations the basic constitutive systems of equations of the problem are converted to dimensionless form. The problem has been solved by using a powerful analytic tool HAM (Homotopy Analysis Method) [55–58]. The influences of all the emerging parameters on spray rate, velocity profile as well as temperature and pressure distributions have been demonstrated in figures.

2. Materials and Methods

Basic Equations

A motion of a steady and axi-symmetric sprayed liquid thin film CuO-H_2O nanofluid past a stretching cylinder in two dimensions is considered in the presence of magnetic field. The magnetic field is uniform of intensity B_0 acting in the radial way. The Reynolds number due to magnetic field ($Re_m = \sigma_{nf}\mu_e W_w \delta$, where σ_{nf}, μ_e, W_w and δ stand for the electrical conductivity of the nanofluid, magnetic permeability, surface velocity and thickness of the fluid film respectively) is taken less in magnitude as compared to the applied magnetic field. Ignoring viscous dissipation by taking it very less in amount. When the material stretches, the thickness of the cylinder reduces but the outer radius a of the cylinder remains fixed. An axisymmetric spray in a radial direction having velocity V condenses as a film and is dragged along through the help of outer surface of the cylinder. Also, it is suppose that the base fluid and the nanoparticles are in thermal equilibrium prevailing the absence of slip. Selecting z-axis along the cylinder and r-axis is taken radially, as depicted in Figure 1.

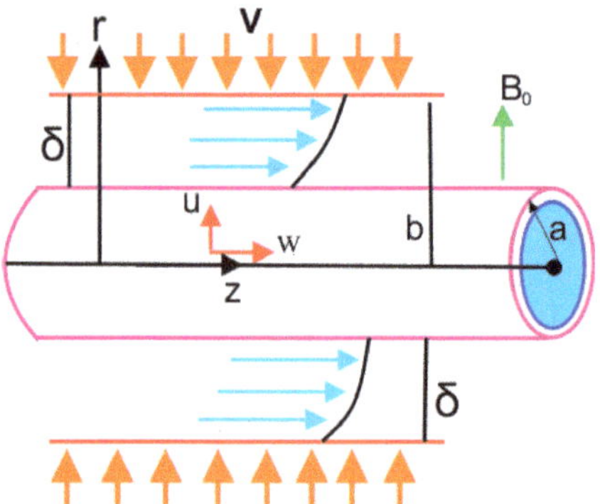

Figure 1. Geometry of the Physical Model.

According to the above statement, in cylindrical coordinates (r, z), the equations of continuity, momentum and energy, governing the film in scalar form are:

$$\frac{\partial(ru)}{\partial r} + \frac{\partial(rw)}{\partial z} = 0, \tag{1}$$

$$\rho_{nf}\left(u\frac{\partial u}{\partial r} + w\frac{\partial u}{\partial z}\right) = -\frac{\partial P}{\partial r} + \mu_{nf}\left(\frac{1}{r}\frac{\partial u}{\partial r} - \frac{u}{r^2} + \frac{\partial^2 u}{\partial r^2}\right), \tag{2}$$

$$\rho_{nf}\left(u\frac{\partial w}{\partial r} + w\frac{\partial w}{\partial z}\right) = \mu_{nf}\left(\frac{1}{r}\frac{\partial w}{\partial r} + \frac{\partial^2 w}{\partial r^2}\right) - \sigma_{nf}B_0^2 w, \tag{3}$$

$$\rho_{nf}\left(u\frac{\partial T}{\partial r} + w\frac{\partial T}{\partial z}\right) = \frac{k_{nf}}{(\rho C_P)_{nf}}\left(\frac{1}{r}\frac{\partial T}{\partial r} + \frac{\partial^2 T}{\partial r^2}\right), \tag{4}$$

where the components of velocity are $u(r, z)$ and $w(r, z)$, the magnetic induction is $B = \left(B_0, 0, 0\right)$, P expresses the pressure and T is the fluid temperature. Further, $\rho_{nf}, \mu_{nf}, (C_P)_{nf}$ and k_{nf} are the density, viscosity, heat capacitance and thermal conductivity of the nanofluid respectively. The subscript "nf" is used for the nanofluid. The effective density (ρ_{nf}) and the heat capacitance $(\rho C_P)_{nf}$ which are the relation between base fluid and nanoparticles are defined as

$$\rho_{nf} = \rho_f(1-\phi) + \rho_s\phi_s, \tag{5}$$

$$(\rho C_P)_{nf} = (\rho C_P)_f(1-\phi) + (\rho C_P)_s\phi, \tag{6}$$

where ϕ describes the solid volume fraction and the subscript "s" is used for nanosolid particles. The boundary conditions are

$$u = U_w, \quad w = W_w, \quad T = T_w, \quad at \quad r = a, \tag{7}$$

$$\frac{\partial w}{\partial r} = \frac{\partial \delta}{\partial r} = \frac{\partial T}{\partial r} = 0, \quad u = \frac{d\delta}{dz} \quad at \quad r = b, \tag{8}$$

where $U_w = -ca$, $W_w = 2cz$, $T_w(z) = T_b - T_{ref}\left[\frac{cz^2}{\nu_{nf}}\right]$.

The subscript "w" is used for condition at the surface, $c > 0$ is the stretching rate, a is the outer radius of the cylinder and b is the outer radius of the film. $\nu_{nf} = \frac{\mu_{nf}}{\rho_{nf}}$ is the kinematic viscosity of the nanofluid, T_b is the temperature at the outer radius of the film surface and T_{ref} is the reference temperature such that $0 \le T_{ref} \le T_b$. The argument $\frac{cz^2}{\nu_{nf}}$ in $T_{ref}\left[\frac{cz^2}{\nu_{nf}}\right]$ shows the stretching velocity along z-axis [31–38].

Introducing the transformations for nondimensional variables f, θ and similarity variable ζ as

$$u = -ca\frac{f(\zeta)}{\sqrt{\zeta}}, \quad w = 2czf'(\zeta), \quad T(z) = T_b - T_{ref}\left[\frac{cz^2}{\nu_{nf}}\right]\theta(\zeta), \quad \zeta = \left(\frac{r}{a}\right)^2. \tag{9}$$

For the outer radius b of the film,

$$\zeta = \left(\frac{b}{a}\right)^2 = \beta_1, \tag{10}$$

where β_1 is the nondimensional film thickness parameter. Equation (9) automatically satisfies mass conservation Equation (1). Utilizing Equations (9)–(10) in Equations (3), (4), (7) and (8) yield the following four Equations (11)–(14)

$$\zeta f''' + f'' - ReB_1(1-\phi)^{2.5}\left(f'^2 - ff''\right) - Mf' = 0, \tag{11}$$

$$\zeta\theta'' + \theta' - RePr\left(\frac{B_2}{B_3}\right)\left(2f'\theta - f\theta'\right) = 0, \tag{12}$$

$$f = f' = \theta = 1 \quad at \quad \zeta = 1, \tag{13}$$

$$f'' = \theta' = 0 \quad at \quad \zeta = \beta_1, \tag{14}$$

where prime ($'$) is the derivative with respect to ζ. $Re = \dfrac{ca^2}{2\nu_{nf}}$, $Pr = \dfrac{\mu_{nf}(\rho C_P)_f}{\rho_f K_f}$ and $M = \dfrac{a^2 B_0^2 \sigma_{nf}}{4\mu_{nf}}$ are respectively the stretching Reynolds number, Prandtl number and magnetic field parameter. The parameters B_1, B_2, B_3 are defined as

$$B_1 = \frac{\rho_s}{\rho_f}\phi(1-\phi), \tag{15}$$

$$B_2 = \frac{(\rho C_P)_s}{(\rho C_P)_f}\phi(1-\phi), \tag{16}$$

$$B_3 = \frac{k_{nf}}{k_f} = \frac{2k_f + k_s - 2(k_f - k_s)\phi}{2k_f + k_s + (k_f - k_s)\phi}, \tag{17}$$

k_f and k_s denote the thermal conductivity of the base fluid and nanosolid particles respectively. A subscript "f" is used for base fluid.

Evaluating Equation (2) for the pressure P

$$\frac{P}{\rho_{nf}} = \frac{P_\infty}{\rho_{nf}} - \frac{(ca)^2 f^2}{2\zeta} - 2c\nu_{nf} f', \tag{18}$$

$\Longrightarrow$

$$\frac{P - P_\infty}{c\mu_{nf}} = -\frac{ReB_1(1-\phi)^{2.5} f^2}{\zeta} - 2f'. \tag{19}$$

The shear stress on the outer film surface is zero, i.e.,

$$f'(\beta_1) = 0. \tag{20}$$

On the cylinder the shear stress is

$$\tau = \frac{\rho_{nf}\nu_{nf}4czf''(1)}{a} = \frac{4c\mu_{nf}zf''(1)}{a}. \tag{21}$$

Estimation of skin friction coefficient (C_f) and heat transfer coefficient (Nu) which are very important through the industrial application point of view are also calculated in this study. The equation defining the skin friction (C_f) is

$$C_f = \frac{2\tau_w}{\rho w_w^2}, \tag{22}$$

where

$$\tau_w = \mu_{nf}\left(\frac{\partial w}{\partial r}\right)_{r=a}, \tag{23}$$

i.e.,

$$\tau_w = \frac{\rho_{nf}\nu_{nf}4czf''(1)}{a} = \frac{4c\mu_{nf}zf''(1)}{a}. \tag{24}$$

Using Equation (22), one obtains:

$$C_f = |\frac{1}{A_1(1-\phi)^{2.5}}f''(1)|, \tag{25}$$

The equation defining the Nusselt number (Nu) is:

$$Nu = \frac{aq_w}{k(T_w - T_b)}, \tag{26}$$

where q_w is the surface heat flux and is given by

$$q_w = -k_{nf}\left(\frac{\partial T}{\partial r}\right)_{r=a}. \tag{27}$$

$\Longrightarrow$

$$q_w = -2k_{nf}\frac{(T_w - T_b)}{a}\theta'(1). \tag{28}$$

Using Equation (26), one obtains

$$Nu = -2\frac{k_{nf}}{k_f}\theta'(1). \tag{29}$$

The deposition velocity V in terms of film thickness β_1 is given by

$$-ca\frac{f(\beta_1)}{\sqrt{\beta_1}} = -V. \tag{30}$$

Mass flux m_1 is another interesting quantity which is in connection with the deposition per axial length is

$$m_1 = V2\pi b. \tag{31}$$

The normalized mass flux m_2 is

$$m_2 = \frac{m_1}{2\pi a^2 c} = \frac{m_1}{4\pi\nu Re} = f(\beta_1). \tag{32}$$

The Brownian motion bears a leading role in the effective thermal conductivity. The effective thermal conductivity is made of particle's conventional static and a Brownian motion part and describes the influences of size, volume fraction, temperature dependence, the type of particle and the base fluid combinations [47].

$$k_{eff} = k_{static} + k_{Brownian}, \tag{33}$$

$$\frac{k_{static}}{k_f} = 1 + \frac{3\left(\frac{k_p}{k_f} - 1\right)\phi}{\left(\frac{k_p}{k_f} + 2\right) - \left(\frac{k_p}{k_f} - 1\right)\phi}, \tag{34}$$

where k_{static} denotes the static thermal conductivity which depends on the Maxwell classical correlation. The enhanced thermal conductivity component generated due to micro-scale convective transfer of heat of a particles's Brownian motion and influenced by ambient fluid flow is obtained through simulating Stoke's motion with a nanoparticle. Utilization of empirical functions (β and f) [48] combines the interaction among nanoprticales with the temperature influence in the model, giving to

$$k_{Brownian} = 5 \times 10^4 \beta\phi\rho_f C_{p,f}\sqrt{\frac{K_b T}{\rho_p d_p}} f(T,\phi). \tag{35}$$

The thermal interfacial resistance (Kapitza resistance) [49,50] exists with the adjacent layers of the both types of materials; the thin barrier layer has a prominent activity in weakening the effective thermal conductivity of the nanoparticle. Li [51] revisited the work of Koo and Kleinstreuer [52] composing a new g'-function by using β and f functions having the effects of particle diameter, temperature and volume fraction. The empirical g'-function is reliant to the used nanofluid [52]. Also, by proposing $R_f = 4 \times 10^{-8}$ km^2/W as a thermal interfacial resistance, the actual k_p in Equation (34) changed to another one, which is $k_{p,eff}$ and has the form

$$R_f + \frac{d_p}{k_p} = \frac{d_p}{k_{p,eff}}, \tag{36}$$

For Al_2O_3-H_2O and CuO-H_2O, the function is

$$g'\left(T,\phi,d_p\right) = \left(c_1 + c_2 ln(d_p) + c_3 ln(\phi) + c_4 ln(\phi) ln(d_p) + c_5 ln(d_p)^2\right) + \left(c_6 + c_7 ln(d_p) + c_8 ln(\phi) + c_9 ln(\phi) ln(d_p) + c_{10} ln(d_p)^2\right) ln(T), \tag{37}$$

where the coefficients c_i (i = 1, 2, 3, ..., 10) are dependent on the type of nanoparticles. With c_i (i = 1, 2, 3, ..., 10), Al_2O_3-H_2O and CuO-H_2O keep an R^2 of 96 % and 98 %, respectively [52] (see Tables 1 and 2). The *KKL*(Koo-Kleinstreuer-Li) correlation becomes

$$k_{Brownian} = 5 \times 10^4 \phi\rho_f C_{p,g}\sqrt{\frac{K_b T}{\rho_p d_p}} g'(T,\phi,d_p). \tag{38}$$

Koo and Kleinstreuer [47] also analyzed the laminar motion of nanofluid in micro heat-sinks by utilizing the effective nanofluid thermal conductivity model. For the effective viscosity because of micro mixing in dispersions, they proposed:

$$\mu_{eff} = \mu_{static} + \mu_{Brownian} = \mu_{static} + \frac{k_{Brownian}}{k_f} \times \frac{\mu_f}{(Pr)_f}, \tag{39}$$

where $\mu_{static} = \dfrac{\mu_f}{(1-\phi)^{2.5}}$ is the nanofluid viscosity utilized in many studies [53,54].

Table 1. Thermophysical quantities of water and nanoparticles [47].

Thermophysical Quantities	Pure Water	Al_2O_3	CuO
$\rho(Kg/m^3)$	997.1	3970	6500
$C_p(J/KgK)$	4179	765	540
$K(W/mk)$	0.613	25	18
$d_p(nm)$	-	47	29
$\sigma(\Omega m)^{-1}$	0.05	10^{-12}	10^{-10}

Table 2. Coefficients values of the Al_2O_3-H_2O and CuO-H_2O [47].

Coefficients Values	Al_2O_3-H_2O	CuO-H_2O
c_1	52.813488759	−26.593310846
c_2	6.115637295	−0.403818333
c_3	0.6955745084	−33.3516805
c_4	$4.17455552786 \times 10^{-2}$	1.915825591
c_5	0.176919300241	$6.42185846658 \times 10^{-2}$
c_6	−298.19819084	48.40336955
c_7	−34.532716906	−9.787756683
c_8	−3.9225289283	190.245610009
c_9	−0.2354329626	10.9285386565
c_{10}	−0.999063481	−0.72009983664

3. Solution of the Problem by HAM

Using the initial approximate values and auxiliary linear operators for the velocity and temperature fields in the following form

$$f_0(\zeta) = \zeta, \qquad \theta_0(\zeta) = 1, \tag{40}$$

$$\mathbf{L}_f = f'', \qquad \mathbf{L}_\theta = \theta'', \tag{41}$$

having the properties of operators

$$\mathbf{L}_f[C_1 + C_2\zeta + C_3\zeta^2] = 0, \quad \mathbf{L}_\theta[C_4 + C_5\zeta] = 0 \tag{42}$$

with constants C_i (i = 1–5).

3.1. Zeroth-Order Deformation Problems

Introducing the nonlinear operators $\aleph_f$ and $\aleph_\theta$ as

$$\aleph_f[f(\zeta,p)] = \zeta\frac{\partial^3 f(\zeta,p)}{\partial\zeta^3} + \frac{\partial^2 f(\zeta,p)}{\partial\zeta^2} - ReB_1(1-\phi)^{2.5}\left[\left(\frac{\partial f(\zeta,p)}{\partial\zeta}\right)^2 - f(\zeta,p)\frac{\partial^2 f(\zeta,p)}{\partial\zeta^2}\right] - M\frac{\partial f(\zeta,p)}{\partial\zeta}, \tag{43}$$

$$\aleph_\theta[f(\zeta,p),\theta(\zeta,p)] = \zeta\frac{\partial^2\theta(\zeta,p)}{\partial\zeta^2} + \frac{\partial\theta(\zeta,p)}{\partial\zeta} - RePr\left(\frac{B_2}{B_3}\right)\left[2\frac{\partial f(\zeta,p)}{\partial\zeta}\theta(\zeta,p) - f(\zeta,p)\frac{\partial\theta(\zeta,p)}{\partial\zeta}\right], \tag{44}$$

where p is an embedding parameter such that $p\in[0, 1]$.

The equations of zeroth-order deformation are prepared as

$$(1-p)\mathbf{L}_f[f(\zeta,p) - f_0(\zeta)] = ph\aleph_f[f(\zeta,p)], \tag{45}$$

$$(1-p)\mathbf{L}_\theta[\theta(\zeta,p) - \theta_0(\zeta)] = ph\aleph_\theta[f(\zeta,p),\theta(\zeta,p)], \tag{46}$$

where h is denoting the auxiliary non-zero parameter. Equation (45) has the boundary conditions

$$f(1,p) = 1, \quad f'(1,p) = 1, \quad f''(\beta_1,p) = 0. \tag{47}$$

The boundary conditions for Equation (46) are

$$\theta(1,p)=1, \qquad \theta'(\beta_1,p)=0. \tag{48}$$

$$p=0\Rightarrow f(\zeta,0)=f_0(\zeta) \quad and \quad p=1\Rightarrow f(\zeta,1)=f(\zeta), \tag{49}$$

$$p=0\Rightarrow \theta(\zeta,0)=\theta_0(\zeta) \quad and \quad p=1\Rightarrow \theta(\zeta,1)=\theta(\zeta), \tag{50}$$

$f(\zeta,p)$ becomes $f_0(\zeta)$ to $f(\zeta)$ when p assumes the values from 0 to 1. Similarly, $\theta(\zeta,p)$ becomes $\theta_0(\zeta)$ to $\theta(\zeta)$ when p assumes the values from 0 to 1. Using Taylor series expansion and Equations (45) and (46), one obtains

$$f(\zeta,p)=f_0(\zeta)+\sum_{m=1}^{\infty} f_m(\zeta)p^m, \quad f_m(\zeta)=\frac{1}{m!}\frac{\partial^m f(\zeta,p)}{\partial p^m}|_{p=0}, \tag{51}$$

$$\theta(\zeta,p)=\theta_0(\zeta)+\sum_{m=1}^{\infty} \theta_m(\zeta)p^m, \quad \theta_m(\zeta)=\frac{1}{m!}\frac{\partial^m \theta(\zeta,p)}{\partial p^m}|_{p=0}. \tag{52}$$

The convergence of the series is sharply relying on h. Suppose h is taken by choice in a such type that the series converges at p = 1, so Equations (51) and (52) result in

$$f(\zeta)=f_0(\zeta)+\sum_{m=1}^{\infty} f_m(\zeta), \tag{53}$$

$$\theta(\zeta)=\theta_0(\zeta)+\sum_{m=1}^{\infty} \theta_m(\zeta). \tag{54}$$

3.2. m-th Order Deformation Problems

By taking m times derivative with respect to p of Equations (45) and (47) then dividing by $m!$ and substituting p = 0, yield the below simplifications

$$\mathbf{L}_f[f_m(\zeta)-\chi_m f_{m-1}(\zeta)]=hR_m^f(\zeta), \tag{55}$$

$$f_m(1)=f_m'(1)=f_m''(\beta_1)=0, \tag{56}$$

$$R_m^f(\zeta)=\zeta f_{m-1}'''+f_{m-1}''-ReB_1(1-\phi)^{2.5}\sum_{k=o}^{m-1}\left[f_{m-1-k}'f_k'-f_{m-1-k}f_k''\right]-Mf_{m-1}'. \tag{57}$$

By taking m times derivative with respect to p of Equatoins (46) and (48), then dividing by $m!$ and substituting p = 0, yield the below simplifications

$$\mathbf{L}_\theta[\theta_m(\zeta)-\chi_m\theta_{m-1}(\zeta)]=hR_m^\theta(\zeta), \tag{58}$$

$$\theta_m(1)=\theta_m'(\beta_1)=0, \tag{59}$$

$$R_m^\theta(\zeta)=\zeta\theta_{m-1}''+\theta_{m-1}'-RePr\left(\frac{B_2}{B_3}\right)\left[2f_{m-1-k}'\theta_k-f_{m-1-k}\theta_k'\right], \tag{60}$$

$$\chi_m=\begin{cases}0, & m\le 1\\ 1, & m>1.\end{cases} \tag{61}$$

If $f_m^*(\zeta)$ and $\theta_m^*(\zeta)$ are the particular solutions, then the general solutions of Equations (55) and (58) in terms of special solutions are represented as follows:

$$f_m(\zeta)=f_m^*(\zeta)+C_1+C_2\zeta+C_3\zeta^2, \tag{62}$$

$$\theta_m(\zeta)=\theta_m^*(\zeta)+C_4+C_5\zeta. \tag{63}$$

4. Results and Discussion

The Equations (11) and (12) with boundary conditions in Equations (13) and (14) are evaluated by the symbolic computer application MATHEMATICA using HAM program and Equation (19) is computed through HAM solution. It is interested to describe the impact of all the parameters on the non-dimensional velocity profile $f(\zeta)$, temperature distribution $\theta(\zeta)$ and pressure distribution $\frac{P-P_\infty}{c\mu_{nf}}(\zeta)$. The rate of spray $m_2(\beta_1)$ is also discussed. The geometry of the problem is shown in Figure 1. Liao [55–57] introduced h curves for the convergence of the series solution to get the acceptable results of the problems. Therefore, the acceptable h-curves for $f(\zeta)$ and $\theta(\zeta)$ are drawn in the ranges $-2.30 \leq h \leq -0.50$ and $-2.50 \leq h \leq 0.50$ in Figures 2 and 3 respectively. The fact is known about the shear stress and rate of transfer of heat that they depend on the types of nanofluids. The type of the nanofluid has a direct relation with cooling and heating techniques. These qualities are achieved easily and quickly by choosing CuO as nanoparticles compared to Al_2O_3 as nanoparticles. In this research CuO-H_2O is used to elaborate the impact of indispensable parameters.

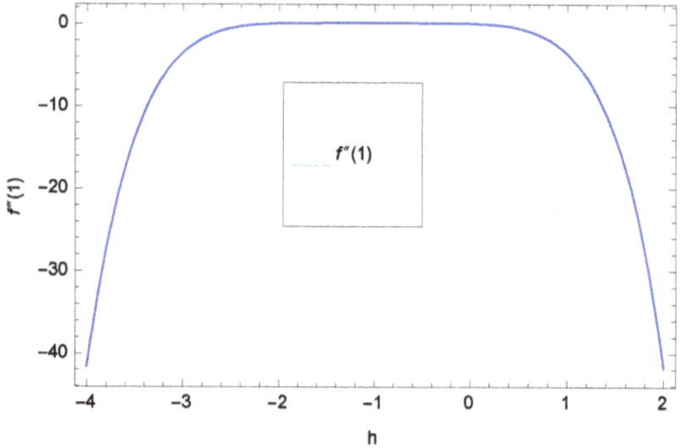

Figure 2. h curve of $f(\zeta)$.

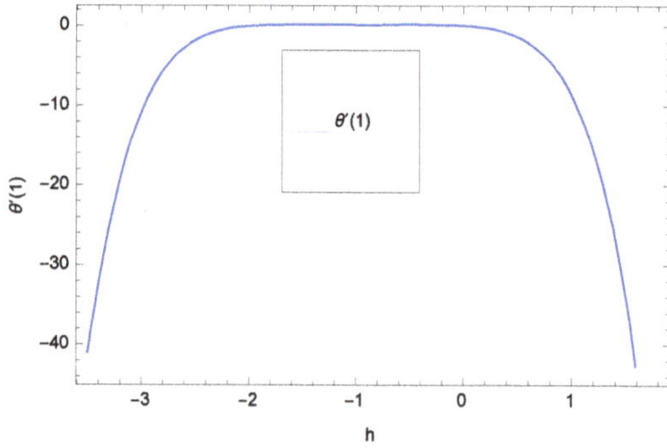

Figure 3. h curve of $\theta(\zeta)$.

5. Discussion

5.1. Velocity Profile

It is elucidated that motion of the fluid slows down near the stretching surface and it enhances far away from stretching cylinder. The salient characters in the film sprayed on a cylinder are the thin film nanofluid parameter β_1 and the stretching Reynolds number *Re*. Concentrating first the contribution of the thin film nanofluid parameter β_1, which has the close relation to the rate of mass spray. It affects the flow behavior completely. From Figure 4, it is evident that the velocity faces retardation with greater quantities of film thickness parameter β_1. The axial velocity $f(\zeta)$ is found to enhance with thinning of the nanoliquid film. With the increase of film thickness the mass of the fluid increases which is difficult to move. Motion automatically stops when thickness of the fluid film is high. Too much effort is required for the movement of the thick film liquid. It is very difficult to make motion in the sea compared to the flow in a pipe. The nanoparticle volume fraction parameter ϕ has a negligible effect on the motion witnessed by Figure 5. Velocity does not change with increasing values of ϕ. Figure 6 demonstrates that the axial velocity $f(\zeta)$ goes to decrease by the large values of Reynolds number *Re*. The reason is that the Reynolds number is the ratio of the inertia force to the viscous force so, when the Reynolds number becomes higher the inertial force overcomes the flow in contrast to viscous forces. Hence for the greater values of Reynolds number *Re* velocity retards and the motion decays gradually to the ambient. The inertial forces are very powerful and they do not allow the liquid atoms/molecules to flow. Strong viscous forces have strong resistance to the motion of the fluid. Boundary layer flow of fluid motion decreases with strong inertial forces. Figure 7 is prepared for the parameter *M* demonstrating magnetic field. As the values of *M* increases on surface during the flow, the motion decreases which results in to retard the motion profiles. Generally, the momentum boundary layer is made thin by greater quantities of *M* since the use of magnetic field to a fluid capable for conduction generates a force of resistivity called Lorentz force. This force is responsible for bringing the retardation in the movement of the fluid. This force decelerates the motion horizontally and overcomes the motion layer related to axial velocity $f(\zeta)$. This characteristic remains up to some heights and then the action is converted to slowing down. The magnetic field parameter keeps an additional influence on the motion. It is clear that magnetic field produces more restriction to the fluid. In fact, magnetic field is in contrast to the transportation.

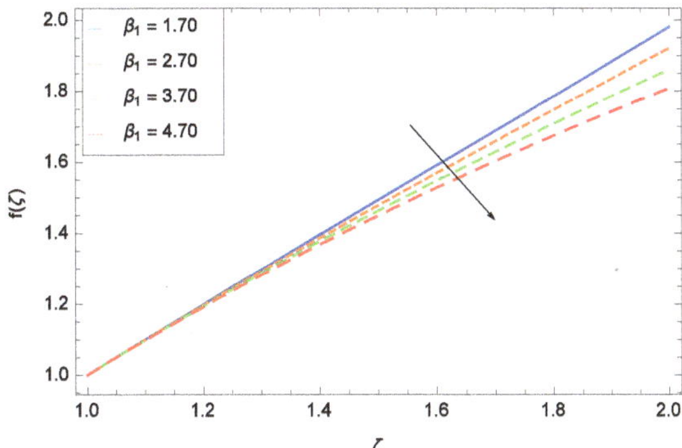

Figure 4. Non-dimensional velocity $f(\zeta)$ sketch for h = - 0.10, M = 0.10, Re = 0.70, ϕ = 0.04, Pr = 6.80 and various values of β_1 (CuO-H_2O).

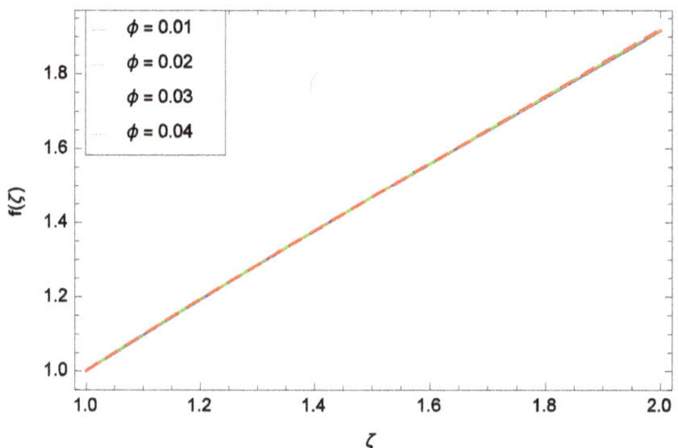

Figure 5. Non-dimensional velocity $f(\zeta)$ sketch for $h = -1.10$, $\beta_1 = 1.70$, $M = 0.10$, $Re = 0.70$, $Pr = 6.80$ and various values of ϕ (CuO-H_2O).

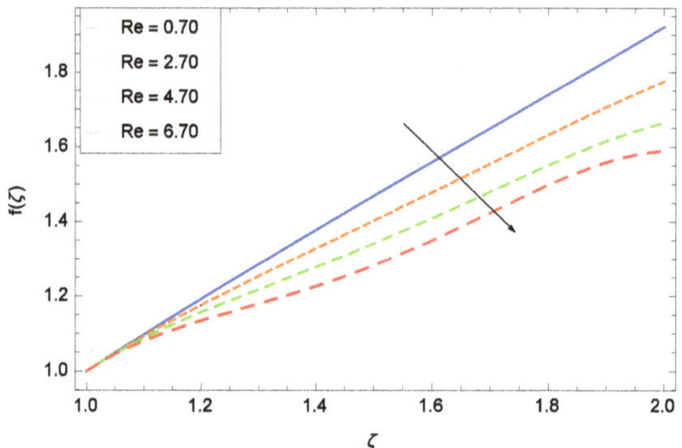

Figure 6. Non-dimensional velocity $f(\zeta)$ sketch for $h = -1.10$, $\beta_1 = 1.70$, $M = 0.10$, $\phi = 0.04$, $Pr = 6.80$ and greater quantities of Re (CuO-H_2O).

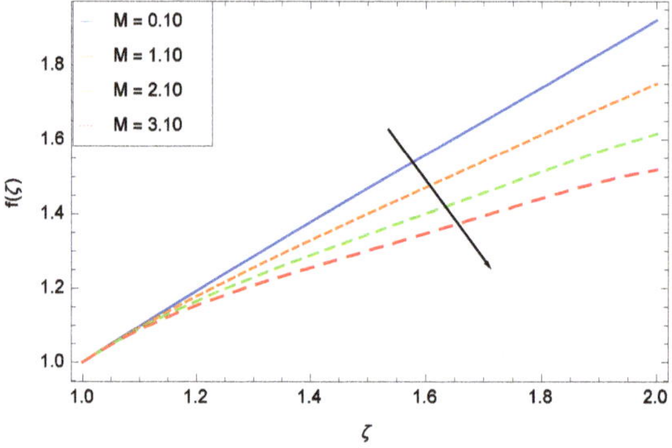

Figure 7. Non-dimensional velocity $f(\zeta)$ sketch for $h = -1.10$, $\beta_1 = 1.70$, $M = 0.10$, $Re = 0.70$, $\phi = 0.04$, $Pr = 6.80$ and greater quantities of M (CuO-H_2O).

5.2. Temperature Profile

The thermal investigation has a prominent application in the cooling of the cylinder. The thin film parameter β_1 has a special role in the temperature distribution. The non-dimensional temperature $\theta(\zeta)$ is high at the surface and it is growing small along the transversely distance within the thermal boundary layer. The dimensional temperature $\theta(\zeta)$ decelerates for greater quantities of film thickness parameter β_1 which is obvious in Figure 8. Transfer of heat is improved for the thinning of the nanofluid film. But in the present case, it is depreciating. The reason is that with the thickness of the fluid film the mass of the fluid is greater which exhaust the amount of temperature. Heat penetrates in the fluid as a result the environment is cool down. Thick film fluid needs more heat as compared to the thin film fluid. The volume fraction parameter ϕ is displayed in Figure 9. No sensitivity occurs in temperature for greater quantities of the volume fraction parameter ϕ. Figure 10 predicts that the temperature $\theta(\zeta)$ depreciates for the big quantities of Reynolds number Re. It is due to the fact that greater Reynolds number signifies that inertial forces exist as the overcoming agents in contrasting the the viscous forces. These inertial forces are very powerful and due to these forces the particles (atoms and molecules) of the fluid remains tightly. Too much heat energy is required in order to break down the bonds between atoms and molecules of the fluids. Due to these forces the boiling point of fluids increase. Figure 11 indicates that temperature distribution enhances with various values of magnetic field parameter M. Due to the application of magnetic the Lorentz force is generated which increases the temperature of the fluid. This force supports and favors the temperature. Since the magnetic field is imposed perpendicularly so with the increasing magnetic field effect the fluid is controlled and bounded. Figure 12 witnesses the influence of Prandtl number Pr describing that $\theta(\zeta)$ decreases with larger values of Pr since for the larger quantities of Pr the thermal boundary layer decreases which displays that effective cooling is achieved quickly for nanofluid. The influence is even more clear for high Prandtl number because the motion layer size is comparatively small. For the greater quantities of Pr the liquid keeps low thermal conductivity which causes thinner thermal boundary layer resulting in the enhancement of the rate of transfer of heat at the surface. Prandtl number defines the ratio of momentum diffusivity to thermal diffusivity. Those liquids which have minimum Prandtl number have good thermal conductivity consequently, possess thick boundary layer structures for diffusing heat.

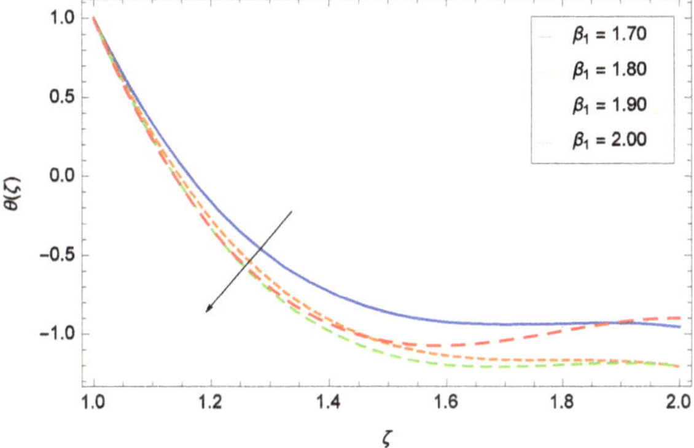

Figure 8. Non-dimensional temperature $\theta(\zeta)$ sketch for $h = -0.10$, $M = 2.00$, $Re = 5.00$, $\phi = 0.04$, $Pr = 6.80$ and various values of β_1 (CuO-H_2O).

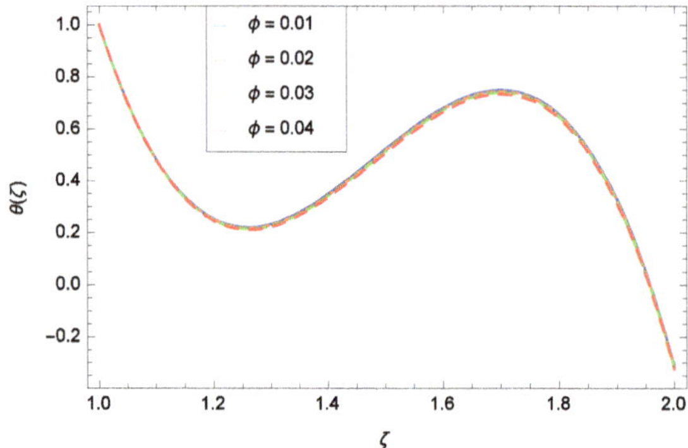

Figure 9. Non-dimensional temperature $\theta(\zeta)$ sketch for $h = -1.10$, $\beta_1 = 1.70$, $M = 0.10$, $Re = 0.70$, $Pr = 6.80$ and various values of ϕ (CuO-H_2O).

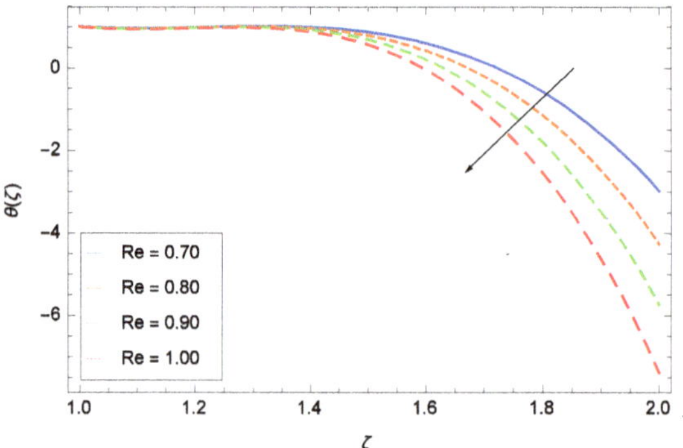

Figure 10. Non-dimensional temperature $\theta(\zeta)$ sketch for $h = -1.10$, $\beta_1 = 1.10$, $\phi = 0.04$, $M = 0.10$, $Pr = 6.80$ and greater quantities of Re (CuO-H_2O).

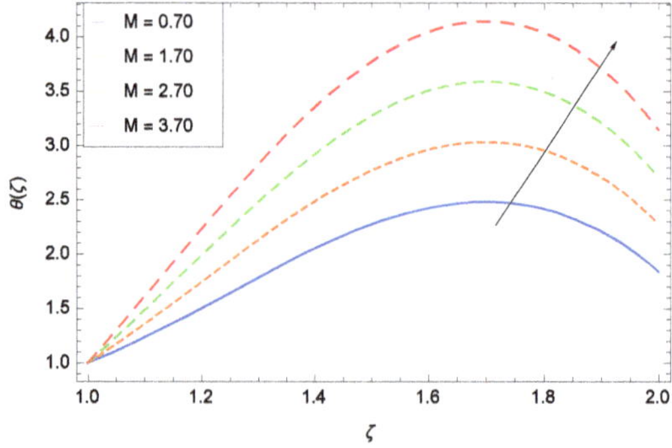

Figure 11. Non-dimensional temperature $\theta(\zeta)$ sketch for $h = -2.50$, $\beta_1 = 1.70$, $\phi = 0.04$, $M = 0.10$, $Pr = 6.80$ and greater quantities of M (CuO-H_2O).

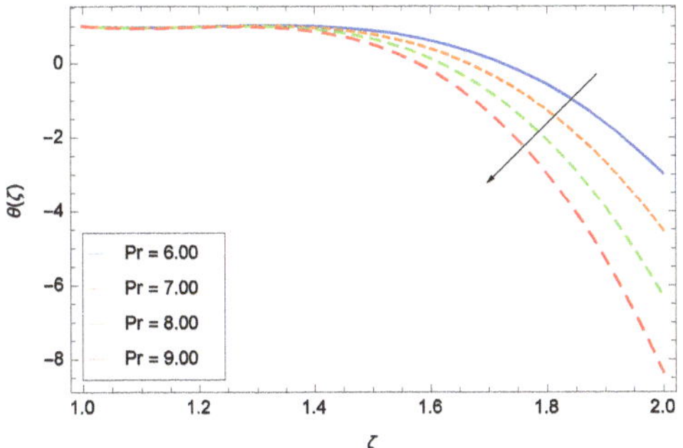

Figure 12. Non-dimensional temperature $\theta(\zeta)$ sketch for $h = -1.10$, $\beta_1 = 1.10$, $M = 0.10$, $\phi = 0.04$, $Re = 0.70$ and various values of Pr (CuO-H_2O).

5.3. Pressure Distribution

This section describes the characteristics of pressure $\frac{P-P_\infty}{c\mu_{nf}}(\zeta)$ in terms of various parameters. Pressure has important contribution in fluid motion and blood flow. Blood circulates in the veins due to pressure. In atmosphere, pressure is inevitable to exist. There are so many machines working due to pressure. Figure 13 indicates that pressure enhances for the greater values of thin film parameter β_1. Pressure distribution is less in magnitude in the wider part of the channel of the motion. The pressure becomes strong with the greater size of film and more power is required to overcome the stress generated due to the thickness of film. The pressure and the film thickness posses a huge force in joint collaboration. This force is very important and is used to help in moving the vehicles/objects over water, in windmills when the fluid flow and objects motion are in the same direction otherwise this force opposes the motion (in case of opposite directions) of the bodies/boats/ships. Figure 14 demonstrates that with the rise of volume fraction parameter ϕ the pressure distribution $\frac{P-P_\infty}{c\mu_{nf}}(\zeta)$ rises because with the addition of nanoparticles the concentration increases, consequently the pressure enhances. When the concentration increases then the fluid becomes thick as a result collision of atoms/molecules increase and exert a great pressure with one another and on the walls of vessel. This characteristic of fluid is very useful for blood flow and for the concentration in medicine. Thick blood runs fast in moving down through pipe due to pressure when injected to the patient. Apart from this, almost all the chemical reactions are made at high pressure. At high pressure, cooking is done easily compared to low pressure. The pressure distribution $\frac{P-P_\infty}{c\mu_{nf}}(\zeta)$ reduces with greater quantities of Reynolds number Re as it is depicted by Figure 15. Pressure automatically goes to minimum value in the wider way of motion due to inertial effects. Due to inertial forces the particles of the fluids are packed tightly and firmly and high pressure is required to make motion from the rest. Pressure has no effect on these attractive forces. High pressure is required to overcome these intermolecular forces. The impact of magnetic parameter M lies in Figure 16. It states that pressure distribution $\frac{P-P_\infty}{c\mu_{nf}}(\zeta)$ is low due to the Lorentz force in the wider channel of the flow. The magnetic is applied perpendicular to the flow of fluid. Lorentz force capture the fluid in the boundary layer. To compete the Lorentz force due to strong magnetic field, pressure must be high in order to make the motion of the fluid.

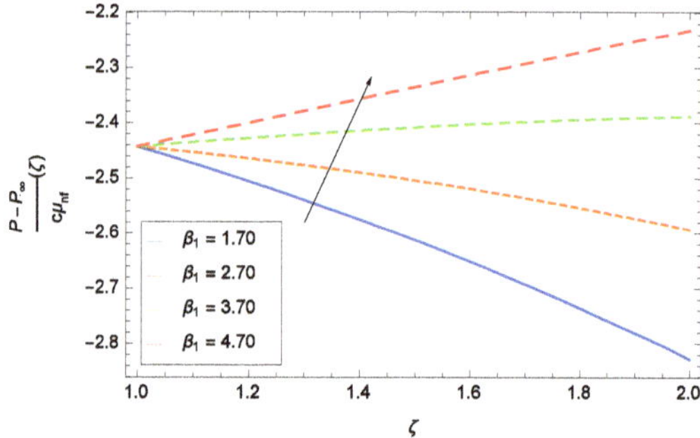

Figure 13. Non-dimensional Pressure $\frac{P-P_\infty}{c\mu_{nf}}(\zeta)$ sketch for $h = -0.10$, $M = 0.10$, $Re = 0.70$, $\phi = 0.04$, $Pr = 6.80$ and various values of β_1 (CuO-H_2O).

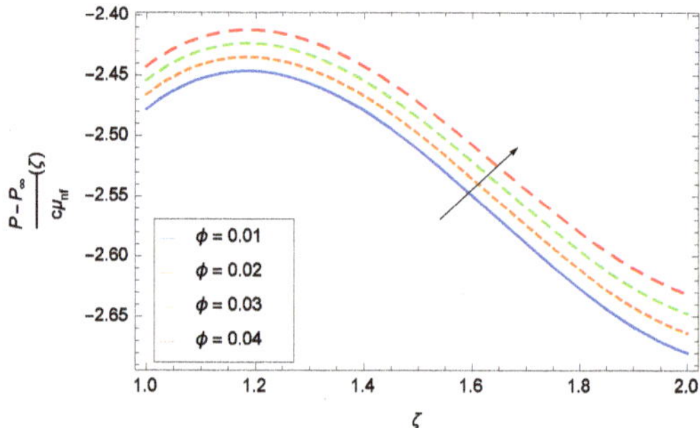

Figure 14. Non-dimensional Pressure $\frac{P-P_\infty}{c\mu_{nf}}(\zeta)$ sketch for $h = -1.10$, $\beta_1 = 1.70$, $M = 0.10$, $Re = 0.70$, $Pr = 6.80$ and various values of ϕ (CuO-H_2O).

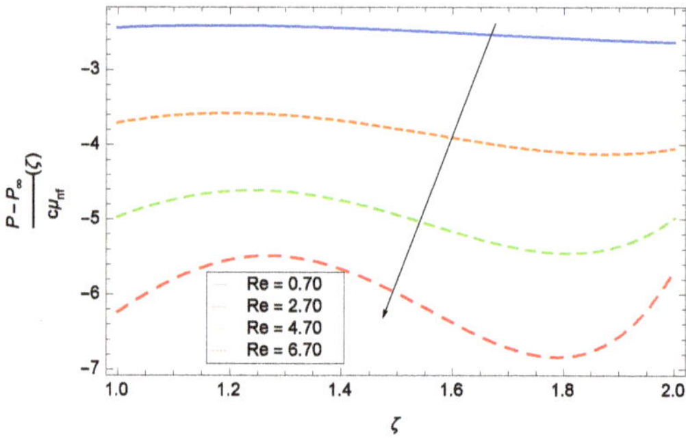

Figure 15. Non-dimensional Pressure $\frac{P-P_\infty}{c\mu_{nf}}(\zeta)$ sketch for $h = -1.10$, $\beta_1 = 1.70$, $M = 0.10$, $\phi = 0.04$, $Pr = 6.80$ and greater quantities of Re (CuO-H_2O).

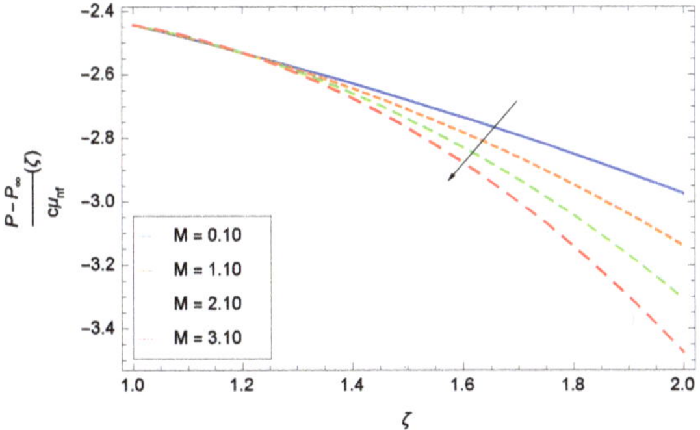

Figure 16. Non-dimensional Pressure $\frac{P-P_\infty}{c\mu_{nf}}(\zeta)$ sketch for $h = -0.10$, $\beta_1 = 1.70$, $M = 0.10$, $Re = 0.70$, $\phi = 0.04$, $Pr = 6.80$ and greater quantities of M (CuO-H_2O).

5.4. Spray Distribution

A stretching cylinder is sprayed by a nanoliquid CuO-H_2O as a coolant and protecting paint or film. Figure 17 expresses the normalized spray rate m_2 which is related functionally to the film size β_1. It is evident that the film size enhances at once by the rate of spray, but it does not occur linearly. If the deposition spray is not uniform, then it is possible to affect the film outer surface.

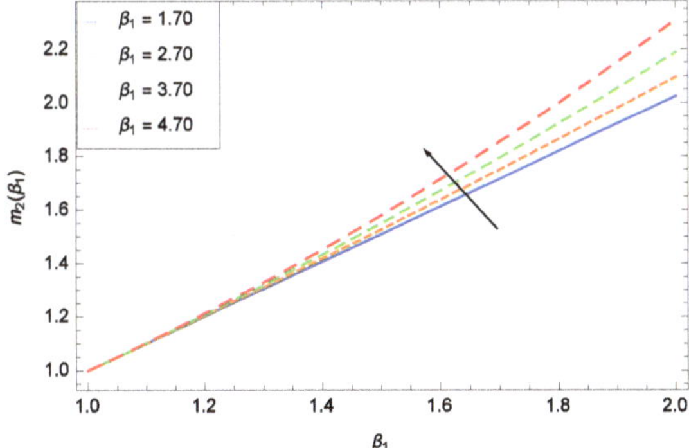

Figure 17. Rate of spray $m_2(\beta_1)$ sketch for $h = 0.10$, $M = 0.10$, $Re = 0.70$, $\phi = 0.04$, $Pr = 6.80$ and various values of β_1(CuO-H_2O).

5.5. Numerical Comparison, Residual Error Sketches and Tables

In order to obtain the accuracy of the achieved results a proper mechanism is adopted. HAM solution is compared with the numerical method solution which shows an excellent agreement with one another. In Figures 18 and 19 the HAM solution graphs of velocity and temperature overlap with the graphs of numerical method solution. To be more authentic the HAM solution, the residual error *Res* graphs are drawn in Figures 20 and 21. Similarly to achieve more precision of HAM solution the comparison Table of HAM solution and numerical method solution is prepared in Table 3. The residual error Table 4 also shows the accuracy of results numerically.

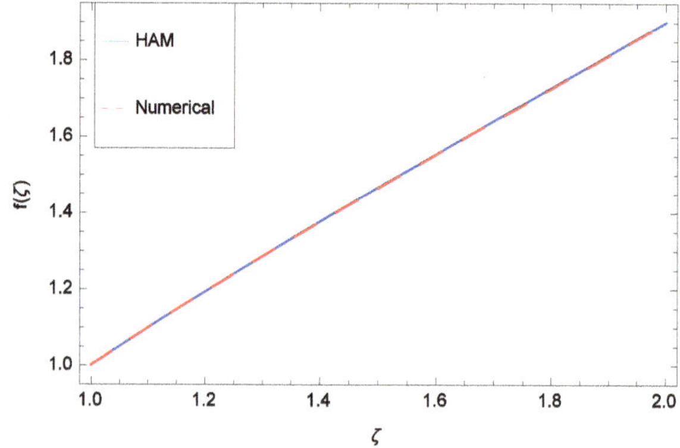

Figure 18. Comparison of the velocity solution of HAM with numerical method solution when $h = -0.55$, $\beta_1 = 2.00$, $M = 0.10$, $Re = 0.70$, $\phi = 0.04$, $Pr = 1.10$ for (CuO-H_2O).

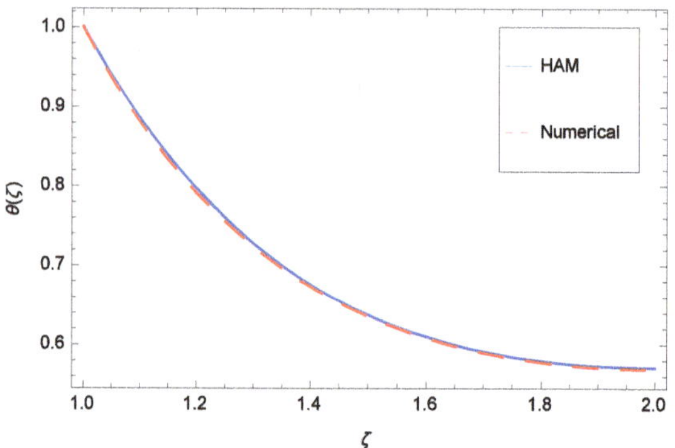

Figure 19. Comparison of the temperature solution of HAM with numerical method solution when $h = -0.55$, $\beta_1 = 2.00$, $M = 0.10$, $Re = 0.70$, $\phi = 0.04$, $Pr = 1.10$ for (CuO-H_2O).

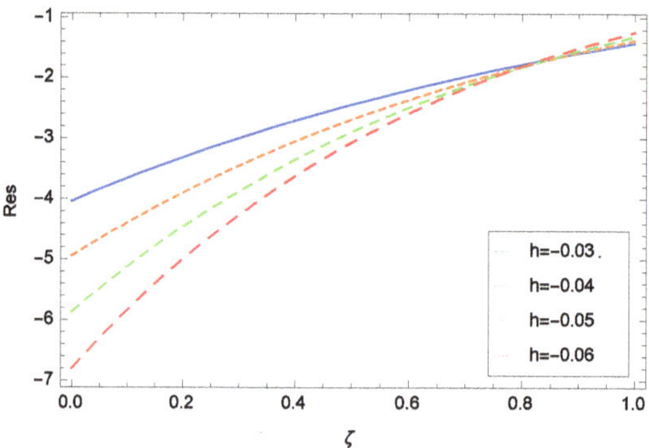

Figure 20. Residual errors sketch of the velocity solution of HAM when $\beta_1 = 2.00$, $M = 0.10$, $Re = 0.70$, $\phi = 0.04$, $Pr = 1.10$ and various values of h for (CuO-H_2O).

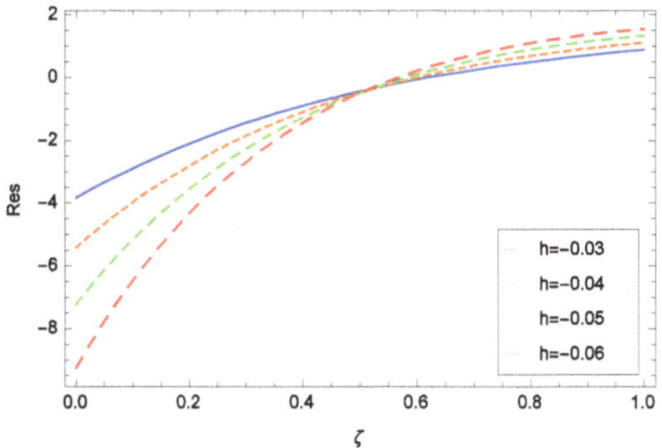

Figure 21. Residual errors sketch of the temperature solution of HAM when $\beta_1 = 2.00$, $M = 0.10$, $Re = 0.70$, $\phi = 0.04$, $Pr = 1.10$ and various values of h for (CuO-H_2O).

Table 3. Comparison of solution of HAM with numerical method solution.

Velocity				Temperature			
ζ	$f(\zeta)$	Numerical Values	Errors	ζ	$\theta(\zeta)$	Numerical Values	Errors
0.0	1.	1.	-4.44089×10^{-16}	0.0	1.0	1.0	-6.30818×10^{-8}
0.1	1.09826	1.09814	0.000124947	0.1	0.885877	0.880696	0.00518123
0.2	1.19347	1.19308	0.000389778	0.2	0.7963263	0.79091	0.00541687
0.3	1.2862	1.28551	0.000692582	0.3	0.7273	0.723303	0.00399767
0.4	1.37694	1.37595	0.000987247	0.4	0.675195	0.672662	0.00253345
0.5	1.46611	1.46485	0.0012588	0.5	0.636847	0.635206	0.00164092
0.6	1.5541	1.55259	0.00150777	0.6	0.609527	0.608145	0.00138164
0.7	1.64121	1.63947	0.0017406	0.7	0.590937	0.589388	0.00154869
0.8	1.72772	1.72575	0.00196426	0.8	0.579203	0.577345	0.00185881
0.9	1.81386	1.81167	0.00218373	0.9	0.572877	0.570793	0.002084
1.0	1.89982	1.89742	0.00240171	1.0	0.570928	0.568784	0.00214339

Table 4. Results achieved by HAM and Residual Errors.

Velocity			Temperature		
ζ	$f(\zeta)$	Residual Errors	ζ	$\theta(\zeta)$	Residual Errors
0.0	1.	−0.64363	0.0	1.0	1.53209
0.1	1.09667	−0.456915	0.1	0.931831	1.64776
0.2	1.18735	−0.29789	0.2	0.875176	1.71672
0.3	1.27298	−0.162766	0.3	0.828685	1.75179
0.4	1.35444	−0.0481967	0.4	0.791164	1.76305
0.5	1.43248	0.0487658	0.5	0.761557	1.75834
0.6	1.50783	0.130716	0.6	0.738932	1.74376
0.7	1.58112	0.199934	0.7	0.722468	1.72407
0.8	1.65293	0.258415	0.8	0.711444	1.70297
0.9	1.72381	0.307909	0.9	0.705227	1.6833
1.0	1.79423	0.349941	1.0	0.703266	1.6673

5.6. Skin Friction Coefficient (C_f), Nusselt Number (Nu) Sketches and Tables Showing Comparison with the Published Experimental Work

It is experimental proved that the shear stress and rate of heat transfer change by using different kinds of nanofluids. From this, it is concluded that the type of nanofluid is most important in the cooling and heating processes. By selecting CuO as nanoparticle the higher Nusselt number and smaller skin friction coefficient can be achieved. Therefore in the present study CuO-H_2O is used to investigate the effects of various parameters. The thermo-physical properties of the nanoliquid are taken to be functions of the volume fraction and the thermal conductivity is modeled based on the effective medium theory. The Prandtl number Pr for the base liquid water is usually around 7. Using the definition of Prandtl number and the thermo-physical properties of water as listed in Table 1 along with $\mu_f = 1 \times 10^{-3}$ Pa s at 20 °C, the Prandtl number of water is evaluated to be $Pr = 6.8173$. This value has been used throughout the computations.

Effects of CuO-H_2O nanofluid on skin friction coefficient (C_f) and Nusselt number (Nu) are demonstrated in Figures 22 and 23. The quantity C_f related to the surface drag is shown as a function of the Reynolds number Re for different values of ϕ. Figure 22 presents the scenario that the skin friction coefficient (C_f) decreases with the increase of nanoparticle volume fraction ϕ. Figure 23 highlights the effect of nanoparticle volume fraction ϕ on Nusselt number Nu as a increasing function of Reynolds number Re. The reason is that the inclusion of nano-sized particles in water like cooling liquids greatly enhances their thermal conductivity thereby causing an increase in the heat transfer rates. The significant variation in Nusselt number against volume fraction in case of CuO-H_2O is observed. The type of nanofluid is a key factor for heat transfer enhancement. The higher values of Nusselt number are obtained by selecting CuO nano-sized particles. Tables 5 and 6 show the comparison of the effects of different kinds of nanoparticles on skin friction coefficient (C_f) and Nusselt number (Nu) in terms of various values of Reynolds number Re against the variation of volume fraction ϕ. Both the Tables numerically show the published experimental results and the present study results which show an excellent correlation.

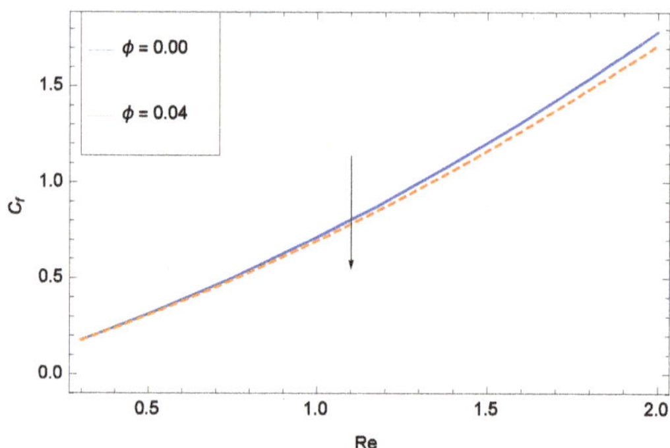

Figure 22. Skin friction coefficient (C_f) when $h = -0.40$, $\beta_1 = 2.00$, $M = 0.10$, $Pr = 6.80$ and various values of ϕ against Reynolds number Re for (CuO-H_2O).

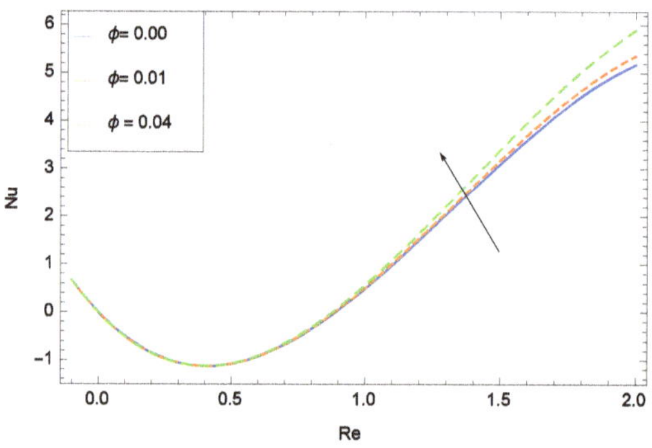

Figure 23. Nusselt number (Nu) when $h = -0.40$, $\beta_1 = 0.10$, $M = 0.10$, $Pr = 6.80$ and various values of ϕ against Reynolds number Re for (CuO-H_2O).

Table 5. Comparison of the effects of different kinds of nanoparticles on skin friction coefficient (C_f) as a function of Reynolds number Re when $\phi = 0.04$.

Parameter	Sheikhoeslami [45]	Sheikhoeslami [45]	Present Study (When $h = -0.20$, $\beta_1 = 0.50$, $M = 0.10$, $Pr = 6.80$)
Re	CuO	Al_2O_3	CuO
0.1	0.679741	0.703897	0.577056
1	1.194617	1.224377	0.215528
2	1.579317	1.615502	0.384453

Table 6. Comparison of the effects of different kinds of nanoparticles on Nusselt number (Nu) as a function of Reynolds number Re when $\phi = 0.04$.

Parameter	Sheikhoeslami [45]	Sheikhoeslami [45]	Present Study (When $h = -0.40$, $\beta_1 = 0.10$, $M = 0.10$, $Pr = 6.80$)
Re	CuO	Al_2O_3	CuO
0.1	1.846686	1.70097	0.227864
1	4.324278	4.113995	2.25099
2	5.996721	5.711652	4.40516

6. Conclusions

This article explores the analytical solution of magnetohydrodynamic thin film CuO-H_2O nanofluid sprayed on a stretching cylinder accompanying transfer of heat. The solution of the problem has been achieved by using analytical technique HAM (Homotopy Analysis Method) for the velocity profile and temperature distribution. Pressure distribution and rate of spray are also investigated. From the given figures, it is clear that the types of nanofluids and various parameters have an effective contribution in the flow, transfer of heat, pressure distribution and rate of spray. The solution has been displayed in the diagrams and the influences of all the parameters included in the problem on the CuO-H_2O nanofluid have been described graphically for checking their effects on velocity profile, temperature distribution and pressure distribution accompanying spray distribution. It is predicted that the model of the problem can be used in technical procedures, cooling/heating techniques and in the modeling of coating processes by employing the constitutive equations of the problem. The findings of the research are the following:

(i) The velocity depreciates for the thin film nanofluid parameter β_1, Reynolds number Re and magnetic field parameter M.
(ii) Temperature diminishes for the thin film parameter β_1, Prandtl number Pr and it enhances for the Reynolds number Re and it elevates for the magnetic field parameter M.
(iii) Volume fraction parameter ϕ has no sensitive effect on velocity and temperature.
(iv) Pressure depreciates for the Reynolds number Re and magnetic field parameter M while it elevates for the thin film parameter β_1, volume fraction parameter ϕ.
(v) Film size β_1 enhances with the spray rate, but nonlinearly.
(vi) Comparisons of HAM solution with the numerical method solution for velocity and temperature are performed and the results are found to be in good agreement.
(vii) The residual errors show the authentication of the present work.
(viii) Skin friction coefficient (C_f) as a function of Reynolds number Re decreases with the increase of nanoparticle volume fraction ϕ.
(ix) Nusselt number Nu as a function of Reynolds number Re increases with the increase of nanoparticle volume fraction ϕ.
(x) The present study shows an excellent agreement with the published experimental work.

Acknowledgments: The authors greatly acknowledge with thanks the Deanship of Scientific Research (DSR) at King Abdulaziz University, Jeddah, Saudi Arabia for technical and financial support. The authors are cordially thankful to the honorable reviewers for their constructive comments to improve the quality of the paper.

Author Contributions: Noor Saeed Khan and Taza Gul modeled and solved the problem. Noor Saeed Khan wrote the paper also. Saeed Islam and Ilyas Khan made the corrections. Aisha M. Alqahtani and Ali Saleh Alshomrani arranged the paper.

Conflicts of Interest: The authors declare no conflict of interest.

Abbreviations

The following abbreviations are used in this manuscript:

MDPI	Multidisciplinary Digital Publishing Institute
DOAJ	Directory of open access journals
TLA	Three letter acronym
LD	linear dichroism

References

1. Choi, S.U.S. Enhancing thermal conductivity of fluids with nanoparticles. In *Developments and Applications of Non-Newtonian Ows*; Siginer, D.A., Wang, H.P., Eds.; ASME: New York, NY, USA, 1995; Volume 66, pp. 99–105.
2. Yu, W.; Xie, H.Q.; Chen, L.F.; Li, Y. Investigation on the thermal transport properties of ethylene glycol-based nanofluids containing copper nanoparticles. *Powder Technol.* **2010**, *197*, 218–221.
3. Yu, W.; Xie, H.Q.; Li, Y.; Chen, L.F. Experimental investigation om the heat transfer properties of Al_2O_3 nanofluids using the mixture of ethylene. *Powder Technol.* **2012**, *230*, 14–19.
4. Chen, L.F.; Yu, W.; Xie, H.Q.; Li, Y. Enhanced thermal conductivity of nanofluids containing Ag/MWNT composites. *Powder Technol.* **2012**, *231*, 18–20.
5. Xie, H.Q.; Chen, L.F.J. Review on the preparation and thermal performances of carbon nanotube contained nanofluids. *Chem. Eng. Data* **2011**, *56*, 1030–1041.
6. Yu, W.; Xie, H.Q.; Chen, W. Experimental investigation on thermal conductivity of nanofluids containing graphene oxide nanosheets. *J. Appl. Phys.* **2010**, *107*, 094317.
7. Xiao, B.; Yang, Y.; Chen, L. Developing a novel form of thermal conductivity of nanofluids with Brownian motion effect by means of fractal geometry. *Powder Technol.* **2013**, doi:10.1016/j.powtec.2013.02.029
8. Cai, J.; Hu, X.; Xiao, B.; Zhou, Y.; Wei, W. Recent developments on fractal-based approaches to nanofluids and nanoparticles aggregation. *Int. J. Heat Mass Transf.* **2017**, *105*, 623–637.
9. Buongiorno, J. Convective transport in nanofluids. *ASME J. Heat Transf.* **2006**, *128*, 240–250.
10. Ellahi, R. The effects of MHD and temperature dependent viscosity on the flow of a non-Newtonian nanofluid in a pipe: Analytical solution. *Appl. Math. Model.* **2013**, *37*, 1451–1457.
11. Khan, W.A.; Pop, I. Boundary layer flow of a nanofluid past a stretching sheet. *Int. J. Heat Mass Transf.* **2010**, *53*, 2477–2483.
12. Mustafa, M.; Hina, S.; Hayat, T.; Alsaedi, A. Influence of wall properties on the peristaltic flow of a nanofluid: Analytic and numerical solutions. *Int. J. Heat Mass Transf.* **2012**, *55*, 4871–4877.
13. Akbar, N.S.; Nadeem, S. Endoscopic effects on peristaltic flow of a nanofluid. *Commun. Theor. Phys.* **2011**, *56*, 761–768.
14. Nowar, K. Peristaltic flow of a nanofluid under the effect of Hall current and porous medium. *Hindawi Publ. Corp. Math. Probl. Eng.* **2014**, *2014*, doi:10.1155/2014/389581.
15. Choi, S.U.S.; Zhang, Z.G.; Yu, W.; Lockwood, F.E.; Grulke, W.A. Anomalously thermal conductivity enhancement in nanotube suspensions. *Appl. Phys. Lett.* **2001**, *79*, 2252–2254.
16. Terekhov, V.I.; Kalinina, S.V.; Lemanov, V.V. The mechanism of heat transfer in nanofluids, State of the Art (Review): Part 1. Synthesis and properties of Nanofluids. *Thermophys. Aeromech.* **2010**, *17*, 1–4.
17. Yu, W.; France, D.M.; Routbort, J.L.; Choi, S.U.S. Review and comparison of nanofluids thermal conductivity and heat transfer enhancements. *Heat Transf. Eng.* **2008**, *29*, 432–460.
18. Hojjat, M.; Etmat, S.G.; Bagheri, R.; Thibault, J. Laminar convective heat transfer of Non-Newtonian nanofluids with constant wall temperature. *Heat Mass Transf.* **2011**, *47*, 203–209.

19. He, Y.; Men, Y.; Liu, X.; Lu, H.; Chen, H.; Ding, Y. Study on forced convective heat transfer of Non-Newtonian nanofluids. *J. Therm. Sci.* **2009**, *18*, 20–26.
20. Polidoiri, G.; Fohanno, S.; Nguyen, C.T. A note on heat transfer modeling of Newtonian nanofluids in laminar free convection. *Int. J. Therm. Sci.* **2007**, *46*, 739–744.
21. Lakshmisha, K.N.; Venkatswaran, S.; Nath, G. Three-Dimensional unsteady flow with heat and mass transfer over a continuous stretching surface. *J. Heat Transf.* **1988**, *110*, 590–595.
22. Wang, C.Y. The three-dimensional unsteady flow due to a stretching flat surface. *Phys. Fluids* **1984**, *27*, 1915–1917.
23. Ahmad, S.; Rohni, A.M.; Pop, I. Blasius and Sakiadis problems in nanofluids. *Acta Mech.* **2011**, *218*, 195–204.
24. Chamkha, A.J.; Aly, A.M.; Al Mudhaf, H. Laminar MHD mixed convection flow of a nanofluid along a stretching permeable surface in the presence of heat generation or absorption effects. *Int. J. Microscale Nanoscale Therm. Fluid Transp. Phenom.* **2011**, 2, 51–70.
25. Kandasamy, R.; Loganathan, P.; Puvi Arasu, P. Scaling group transformation for MHD boundary layer flow of a nanofluid past a vertical stretching surface in the presence of suction and injection. *Nuclear Eng. Des.* **2011**, *241*, 2053–2059.
26. Sakiadis, B.C. Boundary layer behavior on continuous solid surface: Boundary layer on a continuous flat surface. *Am. Inst. Chm. Eng. J.* **1961**, *7*, 213–215.
27. Crane, L.J. Flow past a stretching sheet. *Zeits für Ange Math. Phys.* **1970**, *21*, 645–647.
28. Vajravelu, K.; Rollins, D. Hydromagnetic flow of a second grade fluid over a stretching sheet. *Appl. Maths. Comput.* **2004**, *148*. 783–791.
29. Abu-Nada, E. Effects of variable viscosity and thermal conductivity of Al_2O_3 water nanofluid on heat transfer enhancement in natural convection. *Int. J. Heat Fluid Flow* **2009**, *30*, 679–690.
30. Nasrin, R.; Alim, M.A. Entropy generation by nanofluid with variable thermal conductivity and viscosity in a flat plate solar collector. *Int. J. Eng. Sci. Technol.* **2015**, 7, 80–93.
31. Khan, Y.; Wua, Q.; Faraz, N.; Yildirim, A. The effects of variable viscosity and thermal conductivity on a thin film flow over a shrinking/stretching sheet. *Comput. Math. Appl.* **2011**, *61*, 3391–3399.
32. Khan, N.S.; Gul, T.; Islam, S.; Khan, W.; Khan, I.; Ali, L. Thin film flow of a second-grade fluid in a porous medium past a stretching sheet with heat transfer. *Alex. Eng. J.* **2017**, in press.
33. Khan, N.S.; Gul, T.; Islam, S.; Khan, W. Thermophoresis and thermal radiation with heat and mass transfer in a magnetohydrodynamic thin film second-grade fluid of variable properties past a stretching sheet. *Eur. Phys. J. Plus* **2017**, *132*, doi:10.1140/epjp/i2017-11277-3.
34. Aziz, R.C.; Hashim, I.; Alomari, A.k. Thin film flow and heat transfer on an unsteady stretching sheet with internal heating. *Meccanica* **2011**, *46*, 349–357.
35. Khan, W.; Gul, T.; Idrees, M.; Islam, S.; Dennis, L.C.C. Thin film Williamson nanofluid flow with varying viscosity and thermal conductivity on a time-dependent stretching sheet. *Appl. Sci.* **2016**, *6*, 334.
36. Qasim, M.; Khan, Z.H.; Lopez, R.J.; Khan, W.A. Heat and mass transfer in nanofluid over an unsteady stretching sheet using Buongiorno's model. *Eur. Phys. J. Plus* **2016**, *131*, 1–16.
37. Prashant, G.M.; Jagdish, T.; Abel, M.S. Thin film flow and heat transfer on an unsteady stretching sheet with thermal radiation, internal heating in presence of external magnetic field. *arXiv:1603.3664 physics. flu-dyn.* **2016**, *3*, 1–16.
38. Kumari, M.; Gireesha, B.J.; Gorla, R.S.R. Heat and mass transfer in nanofluid over an unsteady stretching surface. *J. Nanofluids* **2015**, *4*, 1–8.
39. Wang, C.Y. Fluid flow due to a stretching cylinder. *Phys. Fluids* **1988**, *31*, 466–468.
40. Ishak, A.; Nazar, R.; Pop, I. Magnetohydrodynamic (MHD) flow and heat transfer due to a stretching cylinder. *Energ. Convers. Manag.* **2008**, *49*, 3265–3269.
41. Wang, C.Y. Natural convection on a cylinder. *Commun. Nonlinear Sci. Numer. Simul.* **2012**, *17*, 1098–1103.
42. Elbashbeshy, E.M.A.; Emam, T.G.; El-Azab, M.S.; Abdelgaber, K.M. Effect of magnetic field on flow and heat transfer over a stretching cylinder in the presence of a heat source/sink with suction/injection. *J. App. Mech. Eng. I* **2012**, *106*, doi:10.4172/2168-9873.1000106.
43. Ashorynejad, H.R.; Sheikoleslami, M.; Pop, I.; Ganji, D.D. Nanofluid flow and heat transfer due to a stretching cylinder in the presence of magnetic field. *Heat Mass Transf.* **2012**, *49*, 427–436.
44. Rangi, R.R.; Ahmad, N. Boundary layer flow past a stretching cylinder and heat transfer with variable thermal conductivity. *Appl. Math.* **2012**, *3*, 205–209.

45. Sheikholeslami, M. Effect of uniform suction of nanofluid flow and heat transfer over a cylinder. *Braz. Soc. Mech. Sci. Eng.* **2014**, doi:10.1007/40430-014-0242-z.
46. Wang, C.Y. Liquid film sprayed on on a stretching cylinder. *Chem. Eng. Commun.* **2006**, *193*, 869–878.
47. Koo, J.; Kleinstreuer, C. Viscous dissipation effects in micro tubes and micro channels. *Int. J. Heat Mass Transf.* **2004**, *47*, 3159–3169.
48. Koo, J. Computational Nanofluid Flow and Heat Transfer Analysis Applied to Microsystems. Ph.D. Thesis, NC State University, Raleigh, NC, USA, 2004.
49. Prasher, R.S.; Bhattacharya, P.; Phelan, P.E. Thermal conductivity of nano scale colloidal solution. *Phys. Rev. Lett.* **2005**, *94*, 025901.
50. Jang, S.P.; Choi, S.U.S. The role of Brownian motion in the enhanced thermal conductivity of nanofluids. *Appl. Phys. Lett.* **2004**, *84*, 4316–4318.
51. Li, J. Computational Analysis of Nanofluid Flow in Micro Channels with Applications to Micro-Heat Sinks and Bio-MEMS. Ph.D. Thesis, NC State University, Raleigh, NC, USA, 2008.
52. Koo, J.; Kleinstreuer, C. Laminar nanofluid flow in micro-heat-sinks. *Int. J. Heat Mass Transf.* **2005**, *48*, 2652–2661.
53. Brinkman, H.C. The viscosity of concentrated suspensions and solutions. *J. Chem. Phys.* **1952**, *20*, 571.
54. Einstein, A. *Investigation on the Theory of Brownian Motion*; Dover Publications, Inc.: New York, NY, USA, 1956.
55. Liao, S.J. *Homotopy Analysis Method in Non-Linear Differential Equations*; Higher Education Press: Beijing, China; Springer: Berlin/Heidelberg, Germany, 2012.
56. Liao, S.J. On the homotopy analysis method for non-linear problems. *Appl. Math. Comput.* **2004**, *147*, 499–513.
57. Liao, S.J. Homotopy analysis method: A new analytic method for non-linear problems. *Appl. Math. Mech.* **1998**, *19*, 957–962.
58. Gupta, V.G.; Gupta, S. Application of homotopy analysis method for solving nonlinear Cauchy problem. *Surv. Math. Appl.* **2012**, *7*, 105–116.

Article

The Brownian and Thermophoretic Analysis of the Non-Newtonian Williamson Fluid Flow of Thin Film in a Porous Space over an Unstable Stretching Surface

Liaqat Ali [1,2], Saeed Islam [1], Taza Gul [1], Ilyas Khan [3], L. C. C. Dennis [4,*], Waris Khan [5] and Aurangzeb Khan [6]

1 Department of Mathematics, Abdul Wali Khan University, Mardan 23200, KPK, Pakistan; liaqat@cecos.edu.pk (L.A.); saeedislam@awkum.edu.pk (S.I.); tazagulsafi@yahoo.com (T.G.)
2 Department of Electrical Engineering, CECOS University, Peshawer 25000, KPK, Pakistan
3 Department of Mathematics, College of Engineering, Majmaah University, Majmaah 31750, Saudi Arabia; ilyaskhanqau@yahoo.com
4 Department of Fundamental and Applied Sciences, Universiti Teknologi Petronas, Perak 32610, Malaysia
5 Department of Mathematics, Islamia College, Peshawer 25000, KPK, Pakistan; wariskhan758@yahoo.com
6 Department of Physics, Abdul Wali Khan University, Mardan 23200, KPK, Pakistan; akhan@awkum.edu.pk
* Correspondence: dennis.ling@petronas.com.my; Tel.: +60-168-529-529-339

Academic Editor: Richard Yong Qing Fu
Received: 19 February 2017; Accepted: 5 April 2017; Published: 18 April 2017

Abstract: This paper explores Liquid Film Flow of Williamson Fluid over an Unstable Stretching Surface in a Porous Space . The Brownian motion and Thermophoresis effect of the liquid film flow on a stretching sheet have been observed. This research include, to focus on the variation in the thickness of the liquid film in a porous space. The self-similarity variables have been applied to convert the modelled equations into a set of non-linear coupled differential equations. These non-linear differential equations have been treated through an analytical technique known as Homotopy Analysis Method (HAM). The effect of physical non-dimensional parameters like, Eckert Number, Prandtl Number, Porosity Parameter, Brownian Motion Parameter, Unsteadiness Parameter, Schmidt Number, Thermophoresis Parameter, Dimensionless Film Thickness, and Williamson Fluid Constant on the liquid film size are investigated and conferred in this endeavor. The obtained results through HAM are authenticated, from its comparison with numerical (ND-Solve Method). The graphical comparison of these two methods is elaborated. The numerical comparison with absolute errors are also been shown in the tables. The physical and numerical results using h curves for the residuals of the velocity, temperature and concentration profiles are obtained.

Keywords: Thermophoretic effect and Brownian motion, thin film, porous medium, Williamson fluid, unsteady stretching sheet, HAM, ND-solve methods

---

1. Introduction

In the existing literature most of the study is related to Newtonian Fluids and very little attention is paid to the Non-newtonian fluids. Therefore Williamson Fluid has been selected from the class of non-newtonian shear thickening and shear thinning fluids, which has many uses in the field of industry and engineering. The flow of Pseudoplastic Fluids experimentally describe by Williamson [1] with verified results. The analytical study of Williamson Fluid can be found in the investigation of Dapra and Scarpi [2]. Thermophoresis (also Thermomigration, Thermodiffusion, the Soret Effect, or the Ludwig-Soret Effect) is a phenomenon observed in mixtures of mobile particles where the different particle types exhibit different responses, to the force of a temperature gradient. The term Thermophoresis most often applies to aerosol mixtures, but may also commonly refer to the phenomenon in all forms of matter. The term Soret Effect normally applies to liquid mixtures, which behave in different, less well-understood mechanisms than gaseous mixtures. Thermophoresis may not apply to thermomigration in solids, especially multi-phase alloys. The phenomenon is observed at the scale of one millimeter or less. An example that may be observed by the naked eye with good lighting is when the hot rod of an electric heater is surrounded by tobacco smoke, the smoke goes away from the immediate vicinity of the hot rod. As the small particles of air nearest the hot rod are heated, they create a fast flow away from the rod, down the temperature gradient. They have acquired higher kinetic energy with their higher temperature. When they collide with the large, slower-moving particles of the tobacco smoke they push the latter away from the rod. The force that has pushed the smoke particles away from the rod is an example of a Thermophoretic Force. Brownian motion or Pedesis is the random motion of particles suspended in a fluid (a liquid or a gas) resulting from their collision with the fast-moving atoms or molecules in the gas or liquid. Transfer of heat energy play an important role in almost all of the industrial processes. It is used to save energy and reduce processing time in industrial processes. It is also used to raise the thermal rating and increase the working life of equipment. The Liquid Film Flow of Williamson Fluid in a Porous Space over an Unstable Stretching Surface has focused the interest of several researchers because of its many uses in the fields of engineering and industries. The hydrodynamics of a thin liquid film over an unsteady stretching sheet is studied by Wang et al. [3] and Cramer et al. [4] for the first time. The effect of surface mass transfer mixed convection flow is explored by Selim et al. [5]. Das [6] analyzed the impact of thermal radiation on MHD slip flow over a flat plate with variable fluid properties. The effects of radiation and heat transfer on MHD flow of Viscoelastic Liquid and heat transfer over a stretching sheet is studied by Siddeshwar et al. [7]. Nadeem and Hussain [8] solved the problem of flow and heat transfer analysis of Williamson Nanofluid . Hassanien et al. [9] worked on Variable viscosity and thermal conductivity effects on heat transfer by natural convection from a cone and a wedge in porous media. Aziz et al. [10] considered thin film flow and heat transfer on an unsteady stretching sheet with internal heating. Qasim et al. [11] used Buongiorno's model to investigate heat and mass transfer in Nanofluid. Mahesh et al. [12] studied Heat and Mass Transfer in Nanofuid over an unsteady stretching surface.

Ellahi et al. studied Nanofluid over different phenomena mentioned in [13–17]. A detailed data on thin film Williamson Nanofluid Flow with Varying Viscosity and Thermal Conductivity on a Time-Dependent Stretching Sheet is given by Khan et al. [18]. The present research is the study of liquid film flow of Williamson Fluid in a porous medium over an unsteady stretching sheet with the combined effect of Thermophoresis and Brownian motion. The self-similarity variables has been used to convert the modelled equations into a set of non-linear coupled differential equations. The flow of fluid in a porous medium has also a significant role in the field of engineering and especially in Bio-engineering. The purification of liquids through filtration, human lungs, blood filtration are the application of porous media. The flow of fluid in a porous medium on a stretching sheet can be seen in [19,20]. These non-linear differential equations has been tackled through a powerful analytical method known as Homotopy Analysis Method (HAM) [21–28]. The relevant work can also be found in [29–35]. The effect of physical non-dimensional parameters like Porosity Parameter, Unsteadiness Parameter, Prandtl Number, Schmidt Number, and Dimensionless Film thickness on the liquid film size has been investigated and discussed. The results achieved by the HAM and numerical ND-Solve method are compared and presented in the form of figures and tables with absolute error to make understandable for readers.

2. Mathematical Formulation of Model

Suppose a two dimensional incompressible Liaquid Film Flow of Williamson Fluid on a Porous Unsteady Stretching Sheet with thermal radiation, where heat and mass are transferred simultaneously. The coordinate axes are chosen in such away that the x-axis is parallel to the plate while the y-axis is perpendicular to it. The stretching velocity of the sheet is in the direction of the x-axis which have magnitude $U_w = \frac{\alpha x}{1-\gamma t}$, in which $\alpha > 0$ is the stretching velocity constraint and $\gamma \in [0, 1]$.

The temperature $T_w(x,t) = T_0 - T_{ref}(\frac{\alpha x^2}{2v})(1-\gamma t)^{-1.5}$, where T_0 elaborates the temperature at the surface and T_{ref} depicts the reference temperature. Similarly, $C_w(x,t) = C_0 - C_{ref}(\frac{\alpha x^2}{2v})(1-\gamma t)^{-1.5}$ is the volume concentration, where C_0 illustrates the concentration at the surface and C_{ref} shows the reference concentration. The time dependent term $\frac{\alpha x^2}{v(1-\gamma t)}$, indicates the local Reynold number which reliant on the stretching velocity $U_w(x,t)$. Initially the sheet is fixed with the origin and then an external force is applied to stretch the surface in the positive x-axis at the rate $\frac{\alpha}{(1-\gamma t)}$ in time t with velocity $U_w(x,t)$, where $\gamma \in [0,1]$. Now use the above conditions, to get the following equations as:

Continuity Equation,

$$\frac{\partial u}{\partial x} + \frac{\partial v}{\partial y} = 0, \tag{1}$$

Momentum Equation,

$$\frac{\partial u}{\partial t} + u\frac{\partial u}{\partial x} + v\frac{\partial u}{\partial y} = v\frac{\partial^2 u}{\partial y^2} + 2^{0.5}\Gamma v\frac{\partial^2 u}{\partial y^2}\frac{\partial u}{\partial y} - \frac{v\phi}{K}u, \tag{2}$$

Energy Equation,

$$\frac{\partial T}{\partial t} + u\frac{\partial T}{\partial x} + v\frac{\partial T}{\partial y} = \alpha\frac{\partial^2 T}{\partial y^2} + \tau[D_B(\frac{\partial C}{\partial y}\frac{\partial T}{\partial y}) + \frac{D_T}{T_\infty}(\frac{\partial T}{\partial y})^2] + \frac{v}{C_p}[(\frac{\partial u}{\partial y})^2 + 2^{0.5}\Gamma(\frac{\partial u}{\partial y})^3], \tag{3}$$

Concentration Equation,

$$\frac{\partial C}{\partial t} + u\frac{\partial C}{\partial x} + v\frac{\partial C}{\partial y} = D_B\frac{\partial^2 C}{\partial y^2} + \frac{D_T}{T_\infty}(\frac{\partial^2 T}{\partial y^2}), \tag{4}$$

along with the BCs,

$$\begin{aligned} &u = U_w,\ T = T_w,\ v = 0,\ C = C_w,\ y = 0, \\ &\frac{\partial u}{\partial y} = \frac{\partial C}{\partial y} = \frac{\partial T}{\partial y} = 0,\ v = \frac{dh}{dt} = 0,\ y = h(t). \end{aligned} \tag{5}$$

u and v are the flow velocities along x and y axis respectively, the Specific heat at constant pressure is represented by C_p, the Thermal diffusivity of the base fluid is indicated by $\alpha = \frac{k}{(\rho c)p}$, $\Gamma > 0$ is the Time constant, the Fluid density is represented by ρc, $\tau = \frac{(\rho c)p}{(\rho c)f}$ and the local nanoparticle Volume fraction is denoted by C . Also the Thermophoretic diffusion coefficient is indicated by D_T, ρ is the Density, while the Brownian diffusion coefficient is shown by D_B. T is the local Temperature and the Film thickness is denoted by $h(t)$.

Now define the following similarity transformations as:

$$\begin{aligned} &\xi = (\frac{\alpha}{v(1-\gamma t)})^{0.5}y, \\ &\psi(x,y,t) = (\frac{v\alpha}{1-\gamma t})^{0.5}xf(\xi), \\ &T(x,y,t) = T_0 - T_{ref}(\frac{\alpha x^2}{2v})(1-\gamma t)^{-1.5}\theta(\xi), \\ &C(x,y,t) = C_0 - C_{ref}(\frac{\alpha x^2}{2v})(1-\gamma t)^{-1.5}\phi(\xi). \end{aligned} \tag{6}$$

$\psi(x,y)$ is the Stream function which is defined as: $u = \frac{\partial \psi}{\partial y}$, $v = -\frac{\partial \psi}{\partial x}$. β is Non-dimensional film thickness and is described as $\beta = (\frac{\alpha}{v(1-\gamma t)})^{0.5}(h(t))$ [29,30].

Also $\frac{dh}{dt} = -\frac{\gamma}{2}\beta(\frac{\nu}{\alpha})^{0.5}(1-\gamma t)^{-0.5}$. Now put the values in the above equations we get a system of nonlinear coupled boundary value problems as:

$$\begin{gathered}\frac{d^3f(\xi)}{d\xi^3} + \lambda\frac{d^2f(\xi)}{d\xi^2}\frac{d^3f(\xi)}{d\xi^3} + f(\xi)\frac{d^2f(\xi)}{d\xi^2} - (\frac{df(\xi)}{d\xi})^2 - S(\frac{df(\xi)}{d\xi} + \frac{\xi}{2}\frac{d^2f(\xi)}{d\xi^2}) - K_r\frac{df(\xi)}{d\xi} = 0,\\ \frac{d^2\theta(\xi)}{d\xi^2} + Prf(\xi)\frac{d\theta(\xi)}{d\xi} - 2Pr\frac{df(\xi)}{d\xi}\theta(\xi) - Pr(\frac{S}{2}(3\theta(\xi) + \xi\frac{d\theta(\xi)}{d\xi})) +\\ PrNb\frac{d\phi(\xi)}{d\xi}\frac{d\theta(\xi)}{d\xi}) + PrNt(\frac{d\theta(\xi)}{d\xi})^2 + PrE_c((\frac{d^2f(\xi)}{d\xi^2})^2 + \lambda(\frac{d^2f(\xi)}{d\xi^2})^3) = 0,\\ \frac{d^2\phi(\xi)}{d\xi^2} + Sc(\frac{d\phi(\xi)}{d\xi}f(\xi) - 2\frac{df(\xi)}{d\xi}\phi(\xi) - \frac{S}{2}(3\phi(\xi) + \xi\frac{d\phi(\xi)}{d\xi}) + \frac{Nt}{Nb}\frac{d^2\theta(\xi)}{d\xi^2} = 0,\end{gathered} \tag{7}$$

along with transformed boundary conditions,

$$\frac{d^2f(\beta)}{d\xi^2} = 0,\ \frac{df(0)}{d\xi} = 1,\ f(0) = 0,\ f(\beta) = \frac{S\beta}{2},\ \frac{d\theta(\beta)}{d\xi} = 0,\ \theta(0) = 1,\ \frac{d\phi(\beta)}{d\xi} = 0,\ \phi(0) = 1. \tag{8}$$

where

$$\begin{gathered}\lambda = \Gamma U_w(\frac{2\alpha}{\nu(1-\gamma t)})^{0.5}, K_r = \frac{\nu^2\phi(1-\gamma t)}{\alpha K},\\ S = \frac{\gamma}{\alpha}, Pr = \frac{\nu\rho C_p}{k}, E_c = \frac{U_w^2}{C_p(T_w - T_0)},\\ Sc = \frac{\nu}{D_B}, Nb = \tau D_B(C_w - C_\infty), Nt = \frac{\tau D_T(T_w - T_\infty)}{\nu T_\infty}.\end{gathered} \tag{9}$$

3. Materials and Methods

In this section high accuracy of the applied method is applied to system of nonlinear boundary value problems obtained from the new modeled phenomenon. As a result, we see that this method gives best approximation and takes very less time to produce good results.

Solution of Problem

For the solution of system (7) an analytical technique, called Homotopy Analysis Method (HAM) is used. To apply this method we first find the initial guesses $f_0(\xi)$, $\theta_0(\xi)$, $\phi_0(\xi)$ from the following as: Zeroth order problem:

$$\begin{gathered}\frac{d^3f_0(\xi)}{d\xi^3} = 0,\ f_0(0) = 0,\ \frac{df_0(0)}{d\xi} = 1,\ \frac{d^2f_0(\beta)}{d\xi^2} = 0,\\ \frac{d^2\theta_0(\xi)}{d\xi^2} = 0,\ \theta_0(0) = 1,\ \frac{d\theta_0(\beta)}{d\eta} = 0,\\ \frac{d^2\phi_0(\xi)}{d\xi^2} = 0, \phi_0(0) = 1, \frac{d\phi_0(\beta)}{d\xi} = 0,\end{gathered} \tag{10}$$

which gives the solution as

$$f_0(\xi) = \xi,\ \theta_0(\xi) = 1,\ \phi_0(\xi) = 1. \tag{11}$$

The linear operators are chosen as $\psi_f = \frac{d^3f(\xi)}{d\xi^3}$, $\psi_\theta = \frac{d^2\theta(\xi)}{d\xi^2}$ and $\psi_\phi = \frac{d^2\phi(\xi)}{d\xi^2}$ with the following properties

$$\psi_f(C_1 + C_2\xi + C_3\xi^2) = 0, \psi_\theta(C_4 + C_5\xi) = 0, \psi_\phi(C_6 + C_7\xi) = 0, \tag{12}$$

where $C_i, i = 1-7$ are constants. The resultant non-linear operators $\aleph_f$, $\aleph_\theta$ and $\aleph_\phi$ are chosen as:

$$\aleph_f(\xi;\wp) = \frac{d^3f(\xi)}{d\xi^3} + f(\xi)\frac{d^2f(\xi)}{d\xi^2} + \lambda\frac{d^2f(\xi)}{d\xi^2}\frac{d^3f(\xi)}{d\xi^3} - (\frac{df(\xi)}{d\xi})^2 - S(\frac{df(\xi)}{d\xi} + \frac{\xi}{2}\frac{d^2f(\xi)}{d\xi^2}) - K_r\frac{df(\xi)}{d\xi}, \tag{13}$$

$$\aleph_{\theta}[f(\xi;\wp),\theta(\xi;\wp),\phi(\xi;\wp)] = \frac{d^2\theta(\xi)}{d\xi^2} + Prf(\xi)\frac{d\theta(\xi)}{d\xi} - 2Pr\frac{df(\xi)}{d\xi}\theta(\xi) - Pr(\frac{S}{2}(3\theta(\xi) + \xi\frac{d\theta(\xi)}{d\xi})) + PrNb\frac{d\phi(\xi)}{d\xi}\frac{d\theta(\xi)}{d\xi}) + PrNt(\frac{d\theta(\xi)}{d\xi})^2 + E_c((\frac{d^2f(\xi)}{d\xi^2})^2 + \lambda(\frac{d^2f(\xi)}{d\xi^2})^3), \tag{14}$$

$$\aleph_{\phi}[f(\xi;\wp),\theta(\xi;\wp),\phi(\xi;\wp)] = \frac{d^2\phi(\xi)}{d\xi^2} + Sc(\frac{d\phi(\xi)}{d\xi}f(\eta) - 2\frac{df(\xi)}{d\xi}\phi(\xi) - \frac{S}{2}(3\phi(\xi) + \xi\frac{d\phi(\xi)}{d\xi})) + \frac{Nt}{Nb}\frac{d^2\theta(\xi)}{d\xi^2}. \tag{15}$$

The basic idea of HAM is described in [21–28],
Zeroth-order problems:

$$(1-\wp)\psi_f\left[f(\xi;\wp) - f_0(\xi)\right] = \wp\hbar_f\aleph_f[f(\xi;\wp)], \tag{16}$$

$$(1-\wp)\psi_\theta\left[\theta(\xi;\wp) - \theta_0(\xi)\right] = \wp\hbar\aleph_\theta[f(\xi;\wp),\theta(\xi;\wp),\theta(\xi;\wp)], \tag{17}$$

$$(1-\wp)\psi_\phi\left[\phi(\xi;\wp) - \phi_0(\xi)\right] = \wp\hbar_\phi\aleph_\phi[f(\xi;\wp),\theta(\xi;p),\phi(\xi;\wp)]. \tag{18}$$

The equivalent BCs are:

$$f(0;\wp) = 0, \frac{df(0;\wp)}{d\xi} = 1, \frac{d^2f(\xi;\wp)}{d\xi^2} = 0,$$
$$\theta(0;\wp) = 1, \frac{d\theta(0;\wp)}{d\xi} = 0, \phi(0;\wp) = 1, \frac{d\phi(\beta;\wp)}{d\xi} = 0. \tag{19}$$

where $\wp \in [0,1]$ is the imbedding parameter, $\hbar_f, \hbar_\theta$ and $\hbar_\phi$ are used to control the convergence of the solution. When $\wp = 0$ and $\wp = 1$, then:

$$f(\xi;1) = f(\xi), \theta(\xi;1) = \theta(\xi), \phi(\xi;1) = \phi(\xi). \tag{20}$$

Expanding $f(\xi;\wp), \theta(\xi;\wp)$ and $\phi(\xi;\wp)$ in Taylor's series about $\wp = 0$ as:

$$f(\xi) = f_0(\xi) + \sum_{m=0}^{m=\infty} f_m(\xi)\wp^m,$$
$$\theta(\xi) = \theta_0(\xi) + \sum_{m=0}^{m=\infty} \theta_m(\xi)\wp^m,$$
$$\phi(\xi) = \phi_0(\xi) + \sum_{m=0}^{m=\infty} \phi_m(\xi)\wp^m. \tag{21}$$

where

$$f_m(\xi) = \frac{1}{m!}\frac{d^m f(\xi;\wp)}{d\xi^m}\bigg|_{\wp=0},$$
$$\theta_m(\xi) = \frac{1}{m!}\frac{d^m \theta(\xi;\wp)}{d\xi^m}\bigg|_{\wp=0},$$
$$\phi_m(\xi) = \frac{1}{m!}\frac{d^m \phi(\xi;\wp)}{d\xi^m}\bigg|_{\wp=0}. \tag{22}$$

The secondary constraints $\hbar_f, \hbar_\theta$ and $\hbar_\phi$ are selected in such away that the series (21) converges at $\wp = 1$. Use $\wp = 1$ in (21) to get:

$$f(\xi) = f_0(\xi) + \sum_{m=0}^{m=\infty} f_m(\xi),$$
$$\theta(\xi) = \theta_0(\xi) + \sum_{m=0}^{m=\infty} \theta_m(\xi),$$
$$\phi(\xi) = \phi_0(\xi) + \sum_{m=0}^{m=\infty} \phi_m(\xi). \tag{23}$$

The mth -order problem satisfies the following:

$$\begin{aligned}\psi_f\left[f_m(\xi)-\chi_m f_{m-1}(\xi)\right]&=\hbar_f R_m^f(\xi),\\ \psi_\theta\left[\theta_m(\xi)-\chi_m\theta_{m-1}(\xi)\right]&=\hbar_\theta R_m^\theta(\xi),\\ \psi_\phi\left[\varphi_m(\xi)-\chi_m\phi_{m-1}(\xi)\right]&=\hbar_\phi R_m^\phi(\xi).\end{aligned}\tag{24}$$

The boundary conditions for this problem are:

$$\begin{gathered}\frac{d^2f_m(\beta)}{d\xi^2}=0,\ \frac{df_m(0)}{d\xi}=1,\ f_m(0)=0,\ f_m(\beta)=\frac{S\beta}{2},\ \frac{d\theta_m(\beta)}{d\xi}=0,\ \theta_m(0)=1,\\ \frac{d\phi_m(\beta)}{d\xi}=0,\ \phi_m(0)=1.\end{gathered}\tag{25}$$

Here

$$R_m^f(\xi)=\frac{d^3f_{m-1}}{d\xi^3}+\lambda\sum_{k=0}^{m-1}\frac{d^2f_{m-1-k}}{d\xi^2}\frac{d^3f_k}{d\xi^3}+\left[f_{m-1}\frac{d^2f_{m-1}}{d\xi^2}-\sum_{k=0}^{m-1}\frac{df_{m-1-k}}{d\xi}\frac{df_k}{d\xi}-S\left(\frac{df_{m-1}}{d\xi}+\frac{\xi}{2}\frac{d^2f_{m-1}}{d\xi^2}\right)\right]-\mathrm{Kr}\frac{df_{m-1}}{d\xi},\tag{26}$$

$$\begin{aligned}R_m^\theta(\xi)=\frac{d^2\theta_{m-1}}{d\xi^2}+\Pr\left[-\frac{S}{2}\left(3\theta_{m-1}+\xi\frac{d\theta_{m-1}}{d\xi}\right)-2\sum_{k=0}^{m-1}\theta_{m-1-k}\frac{df_k}{d\xi}+\sum_{k=0}^{m-1}f_{m-1-k}\frac{d\theta_k}{d\xi}\right]+\\ E_c\left[\sum_{k=0}^{m-1}\frac{d^2f_{m-1-k}}{d\xi^2}\frac{d^2f_k}{d\xi^2}+\lambda\sum_{k=0}^{m-1}\frac{d^2f_{m-1-k}}{d\xi^2}\sum_{\ell=0}^{k}\frac{d^2f_{k-1}}{d\xi^2}\frac{d^2f_\ell}{d\xi^2}\right]+Nt\sum_{k=0}^{m-1}\frac{d\theta_{m-1-k}}{d\xi}\frac{d\theta_k}{d\xi}\\ +Nb\left(\frac{d\theta_{m-1}}{d\xi}\frac{d\phi_{m-1}}{d\xi}\right),\end{aligned}\tag{27}$$

$$\begin{aligned}R_m^\phi(\xi)=\frac{d^2\phi_{m-1}}{d\xi^2}+Sc\left[\sum_{k=0}^{m-1}f_{m-1-k}\frac{d\phi_j}{d\xi}-2\sum_{k=0}^{m-1}\frac{df_{m-1-k}}{d\xi}\phi_k-\frac{S}{2}\left(3\phi_{m-1}+\xi\frac{d\phi_{m-1}}{d\xi}\right)\right]\\ +\frac{Nt}{Nb}\frac{d^2\theta_{m-1}}{d\xi^2},\end{aligned}\tag{28}$$

where

$$\chi_m=\begin{matrix}0,if\ \wp\le 1\\ 1,if\ \wp>1.\end{matrix}\tag{29}$$

4. Representation of Achieved Results in the Form of Figures and Tables

In this section the results achieved by HAM are shown in the form of figures and tables. The convergence of the series given in (21), $f(\eta)$, $\theta(\eta)$ and $\phi(\eta)$ entirely depend upon the auxiliary parameters $\hbar_f, \hbar_\theta$ and $\hbar_\phi$ which are called $\hbar$-curves. It is selected in such a way that it controls and converges the series solution. The probable selection of $\hbar$ can be found by plotting $\hbar$-curves of $f''(0)$, $\theta'(0)$, $\phi'(0)$. The valid region of $\hbar$ is $-1.5<\hbar_f<-0.5$, $-1.5<\hbar_\theta<-0.5$ and $-1.5<\hbar_\phi<-0.5$. Here $\eta=\xi$ is chosen.

5. Results and Discussion

In this paper the Liquid Film Flow of Non-newtonian Williamson Fluid over an Unstable Stretching Surface in a Porous Space has been investigated. Thermophoresis and Brownian Motion Effect has been countered to the liquid film flow. The governing equations have been transformed through suitable similarity variables into nonlinear coupled differential equations with physical conditions. The solution of the coupled problem has been obtained by using an analytical approach called, HAM. The solution of the coupled problem and fast convergence of this method is mainly focused. This paper has examined the consequences of governing parameters on the transient velocity, temperature, and concentration profiles. Figure 1 illustrates the geometry of model used. Comparisons are carried out between the obtained results and the results achieved by numerical N-Desolve method for velocity, temperature, and concentration profiles (shown in Figures 2–4). The effects of physical parameters appear in the problem, are shown graphically and discussed. Figures 5–7, elaborate the behavior of the non-dimensional unsteady parameter S for velocity, temperature and concentration field during fluid motion in a porous medium past over a Unsteady Stretching Sheet. The unsteady parameter S is inversely related to the stretching constant of the velocity field, where as it is directly related to the stretching constants of the temperature and concentration fields. Therefore, when the values of S are increasing the values of the velocity field are decreasing while the values of the temperature and concentration fields increase. Physically, unsteadiness S produce buoyancy forces in the way of the flow field. These forces resist the fluid flow and therefore, the velocity field falls and the temperature distribution as well as the concentration profile is boosted. The effect of the Williamson Fluid constant λ on the velocity field is illustrated in Figure 8. The velocity is found to reduce when λ is augmented. Because rise in relaxation time causes higher resistance to the fluid flow and as a result reduces the velocity field. Also increase in λ increase the temperature due to increase in resistance to the fluid flow as shown in Figure 9. Non-dimensional porosity parameter Kr have direct relation to viscosity parameters. So a rise in non-dimensional porosity parameter reduces fluid motion as explained in Figure 10. Physically, larger values of Kr generate larger open space and create hurdle to flow and as a result the flow field is retarded. The resistance force produces larger values of Kr which increase the temperature and concentration profiles shown in Figures 11 and 12.

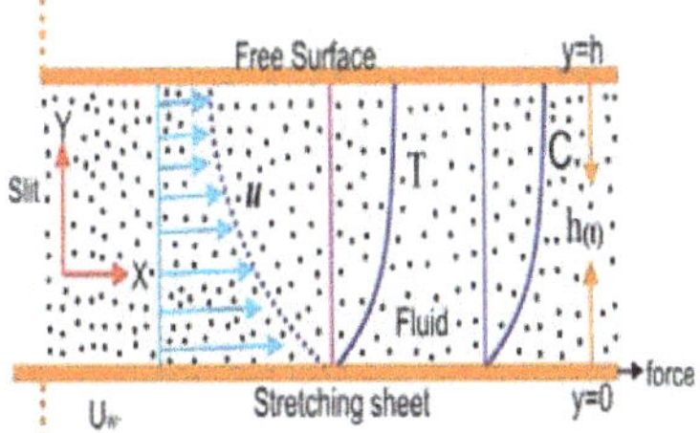

Figure 1. Illustrates the physical geometry of the used model.

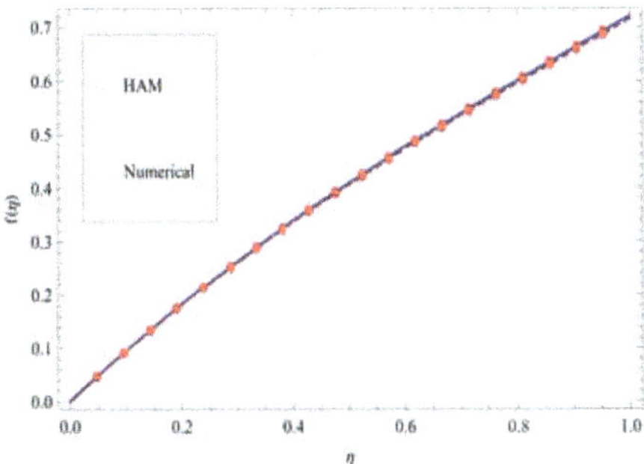

Figure 2. The comparison between HAM and numerical solutions for velocity profile $f(\eta)$, when $h = -0.25, \lambda = 0.9, \mathrm{kr} = 0.9, \mathrm{Pr} = 0.5, \mathrm{Ec} = 0.5, \mathrm{Nb} = 0.5, \mathrm{Nt} = 0.6, \beta = 1$ and $Sc = 0.6$.

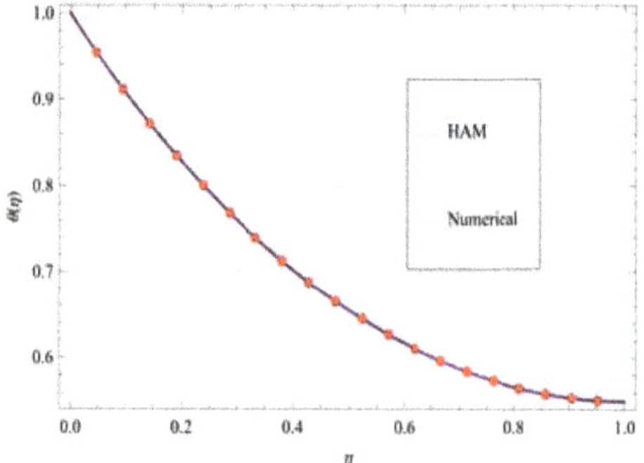

Figure 3. The comparison between HAM and numerical solutions for temperature profile $\theta(\eta)$, when $h = -0.47, \lambda = 0.2, S = 0.2, \mathrm{kr} = 0.2, \mathrm{Pr} = 1, \mathrm{Ec} = 0.6, \mathrm{Nb} = 0.4, \mathrm{Nt} = 0.5, \beta = 1, Sc = 0.5$.

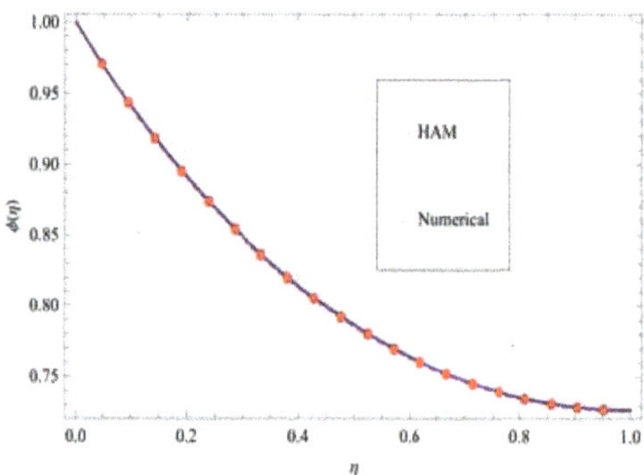

Figure 4. The comparison between HAM and numerical solutions for concentration profile $\phi(\eta)$, when $h = -0.6, \lambda = 0.2, S = 0.2, \mathrm{kr} = 0.2, \mathrm{Pr} = 1, \mathrm{Ec} = 0.6, \mathrm{Nb} = 1, \mathrm{Nt} = 0.1, \beta = 1$ and $Sc = 0.5$.

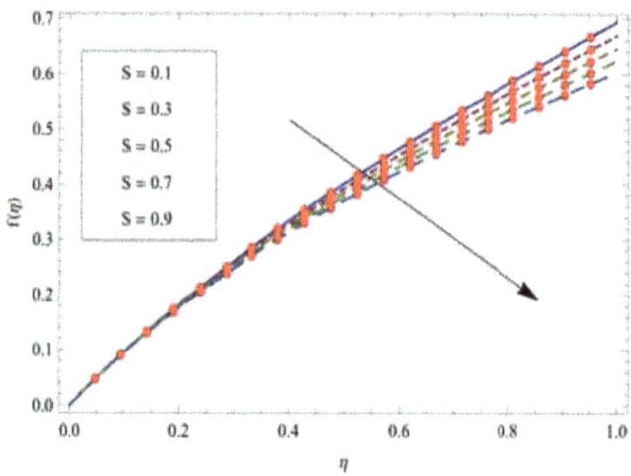

Figure 5. Variations in the Velocity field $f(\eta)$ for various values of S, when $h = -0.25, \lambda = 0.9, \mathrm{kr} = 0.9, \mathrm{Pr} = 0.5, \mathrm{Ec} = 0.5, \mathrm{Nb} = 0.5, \mathrm{Nt} = 0.6, \beta = 1, Sc = 0.6$.

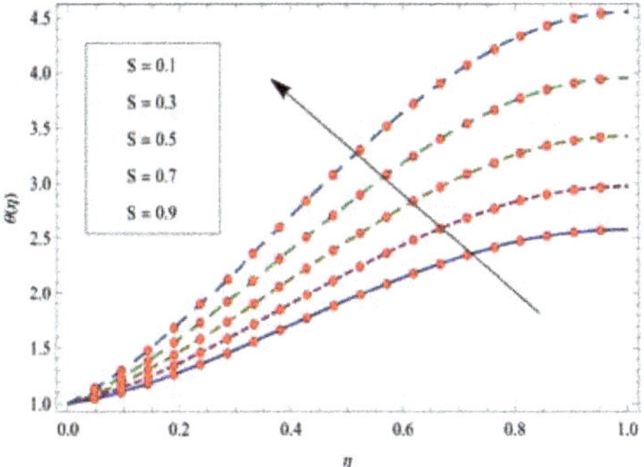

Figure 6. Variations in the Temperature gradient $\theta(\eta)$ for different values of S, when $h = -0.8$, $\lambda = 0.1$, $\mathrm{kr} = 0.5, \mathrm{Pr} = 0.5, \mathrm{Ec} = 0.5, \mathrm{Nb} = 0.5, \mathrm{Nt} = 0.6, \beta = 1$, $Sc = 0.6$.

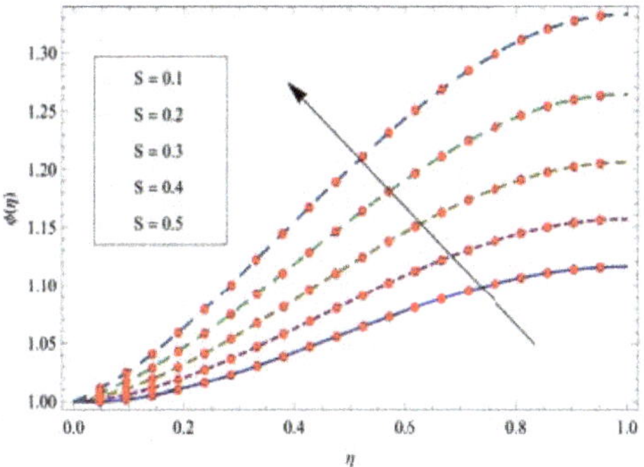

Figure 7. Variations in the Concentration field $\phi(\eta)$ for different values of S, when $h = -0.6$, $\lambda = 0.5$, $\mathrm{kr} = 0.5, \mathrm{Pr} = 0.5, \mathrm{Ec} = 0.5, \mathrm{Nb} = 0.5, \mathrm{Nt} = 0.6, \beta = 1$, $Sc = 0.6$.

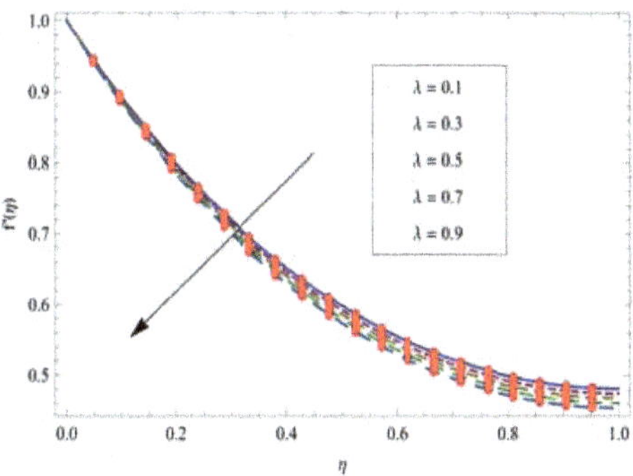

Figure 8. The effect of λ on $f'(\eta)$, when $h = -0.25$, $\mathrm{kr} = 0.7, \mathrm{Pr} = 0.5, \mathrm{Ec} = 0.5, \mathrm{Nb} = 0.5, \mathrm{Nt} = 0.6$, $\beta = 1$, $Sc = 0.6$.

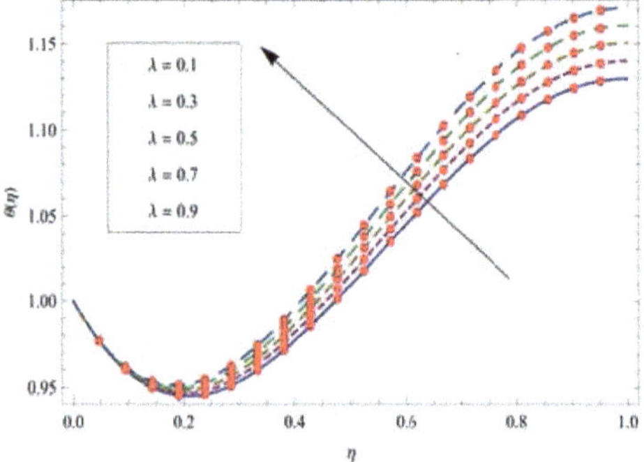

Figure 9. The effect of λ on $\theta(\eta)$, when $h = -0.6$, $S = 0.5, \mathrm{kr} = 0.5, \mathrm{Pr} = 0.5, \mathrm{Ec} = 0.5, \mathrm{Nb} = 0.5$, $\mathrm{Nt} = 0.6, \beta = 1$, $Sc = 0.6$.

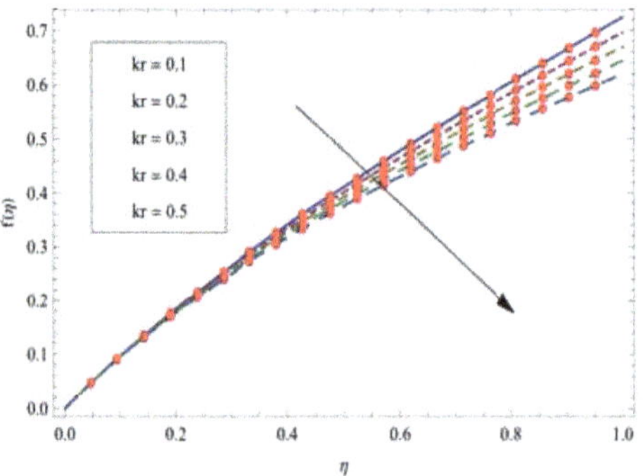

Figure 10. Indicates the effect of Kr on $f(\eta)$ for $h = -0.25$, $S = 0.5, \mathrm{Pr} = 0.5, \lambda = 1, \mathrm{Ec} = 0.5$, $\mathrm{Nb} = 0.5, \mathrm{Nt} = 0.6, \beta = 1$, $Sc = 0.6$.

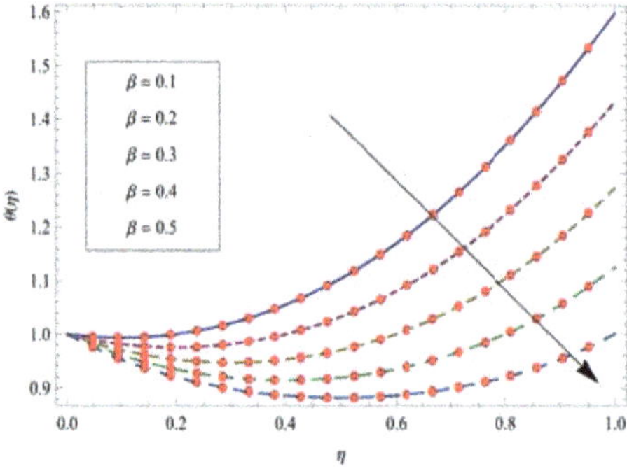

Figure 11. Shows the effect of β on $\theta(\eta)$ for $h = -0.7$, $\lambda = 1, S = 0.5, \mathrm{kr} = 0.5, \mathrm{Pr} = 0.5, \mathrm{Ec} = 0.5$, $\mathrm{Nb} = 0.5, \mathrm{Nt} = 0.6$, and $Sc = 0.6$.

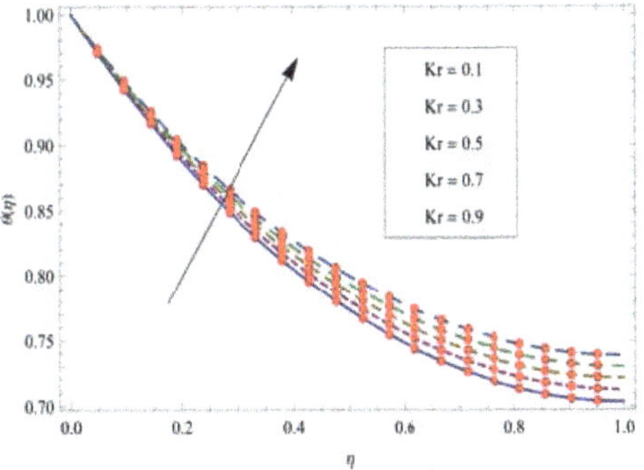

Figure 12. Shows the effect of Kr on $\theta(\eta)$ for $h = -0.25, \lambda = 0.1, S = 0.1, \mathrm{Pr} = 0.5, \mathrm{Ec} = 0.5, \mathrm{Nb} = 0.5$, $\mathrm{Nt} = 0.6$ and $Sc = 0.6$.

The effect of Prandtl number Pr has been shown in the Figure 13, describing that for larger values of Pr decreases the temperature $\theta(\eta)$. The increase in Prandtl number reduces the thermal boundary layer due to which the temperature decreases.The influence of the Schmidt number Sc is depicted in Figures 14 and 15, showing that temperature and concentration fields decrease when the parameter Sc increases because Schmidt number Sc is reciprocal to the molecular diffusivity. It indicates that as the values of of the Eckert number Ec increase the fluid temperature also increases while its converse effect has been observed in the solute concentration illustrated in Figures 16 and 17. Physically, Ec is connected with the viscous dissipation term in the equation of energy, therefore, larger values of Ec should lead to increase the quantity of heat being produced by the shear forces in the fluid and as a result raises the fluid temperature. Figures 18 and 19, illustrate the effects of Brownian motion parameter Nb on the dimensionless temperature and concentration profiles. The fluid temperature increases as the value of increase of Brownian motion parameter increase while converse effect on the solute concentration. The increase in the value of thermophoresis parameter, increase both temperature and concentration as illustrated in Figures 20 and 21. The fluid flow is also falling when the thickness of film is increased. Larger values of thickness β generate the friction force and as a result the flow motion falls down. Increase in the film thickness deliver more fluid in the boundary layer region and cooling effect is produced, which absorbs the heat transfer from the sheet to the fluid and temperature profile drops down. Concentration has vital application in thermal conductivity and chemical reactions. The concentration profile $\phi(\xi)$ is reliant on film size β and increases with larger values of β indicated in Figures 22–24. The h-curves of $f''(0), \theta'(0)$, and $\phi'(0)$ for the 4th-order HAM approximated solution are elaborated in Figures 25–27. Figures 28–31 indicate h curves of the residuals for velocity, temperature and concentration profiles respectively. Table 1 illustrates the symbols used in the manuscript. In Tables 2–4 the results are compared, which are achieved by HAM and Numerical(ND-Solve method) for velocity, temperature and concentration profiles. The residuals gained by HAM are also depicted in Table 5.

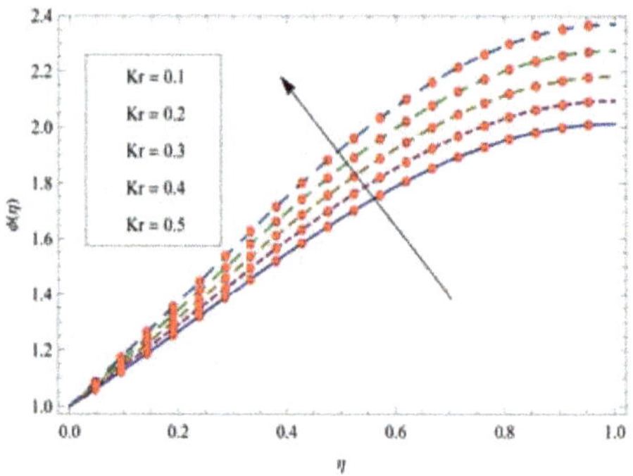

Figure 13. The effect of Kr on $\phi(\eta)$ for $h = -0.9$, $\lambda = 0.5, S = 0.7, \mathrm{Pr} = 0.5, \mathrm{Ec} = 0.5, \mathrm{Nb} = 0.7,$ $\mathrm{Nt} = 0.1, \beta = 1$ and $Sc = 0.5$.

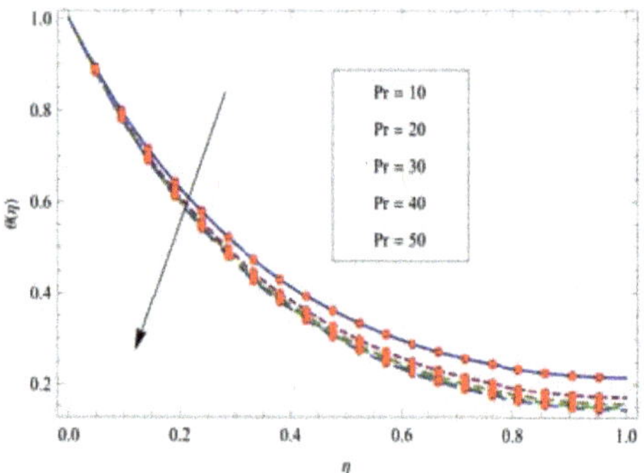

Figure 14. Shows the effect of Pr on $\theta(\eta)$ for $h = -0.7$, $\lambda = 0.7, S = 0.7, \mathrm{kr} = 0.1, \mathrm{Ec} = 0.5, \mathrm{Nb} = 0.5,$ $\mathrm{Nt} = 0.2, \beta = 1$ and $Sc = 0.2$.

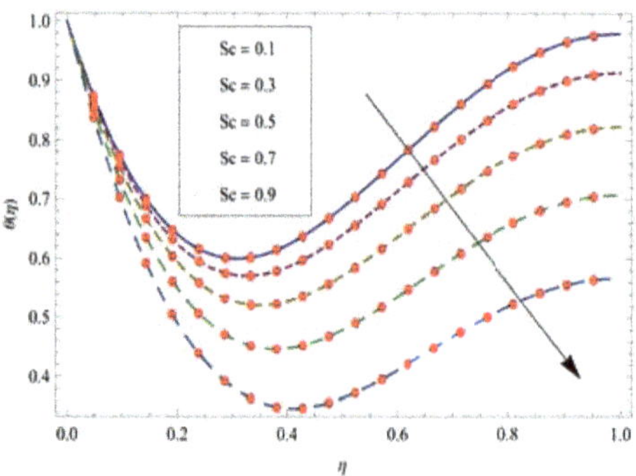

Figure 15. Shows the effect of Sc on $\theta(\eta)$ for $h = -0.9$, $\lambda = 0.7, S = 0.7, \mathrm{kr} = 0.7, \mathrm{Pr} = 30, \mathrm{Ec} = 0.7,$ $\mathrm{Nb} = 0.5, \mathrm{Nt} = 0.7, \beta = 1$ and $Sc = 0.1$.

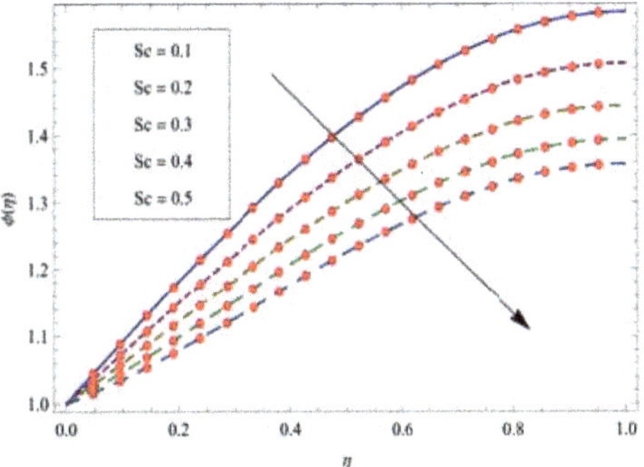

Figure 16. Shows the effect of Sc on $\phi(\eta)$, for $h = -0.6$, $\lambda = 0.5, S = 0.5, \mathrm{kr} = 0.5, \mathrm{Pr} = 0.5, \mathrm{Ec} = 0.5$, $\mathrm{Nb} = 0.5, \mathrm{Nt} = 0.5, \beta = 1$ and $Sc = 0.6$.

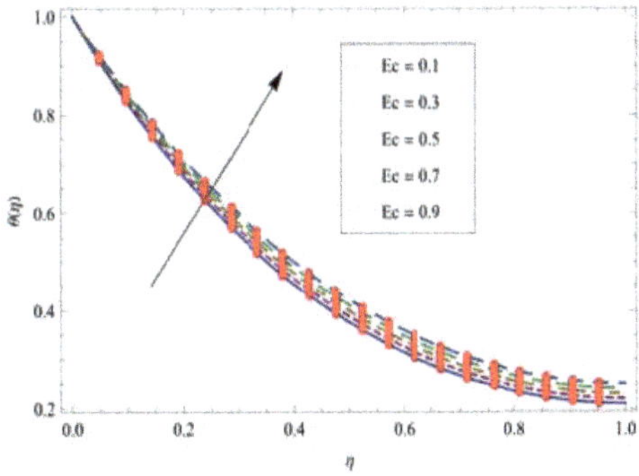

Figure 17. Shows the effect of Ec on $\theta(\eta)$, for $h = -0.6$, $\lambda = 0.1, S = 0.1, \mathrm{kr} = 0.9, Pr = 15, \mathrm{Nb} = 0.5$, $\mathrm{Nt} = 0.6, \beta = 1$ and $Sc = 0.1$.

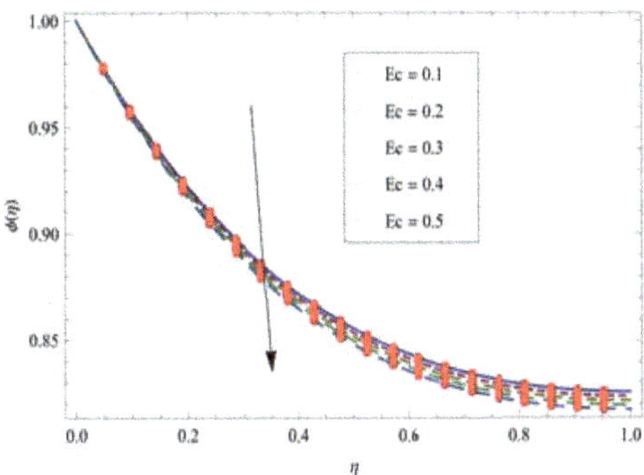

Figure 18. presents the effect of Ec on $\phi(\eta)$ for $h = -0.7$, $\lambda = 0.5, S = 0.7, \mathrm{kr} = 0.2, \mathrm{Pr} = 10, \mathrm{Nb} = 0.7$, $\mathrm{Nt} = 0.1, \beta = 1$ and $Sc = 0.5$.

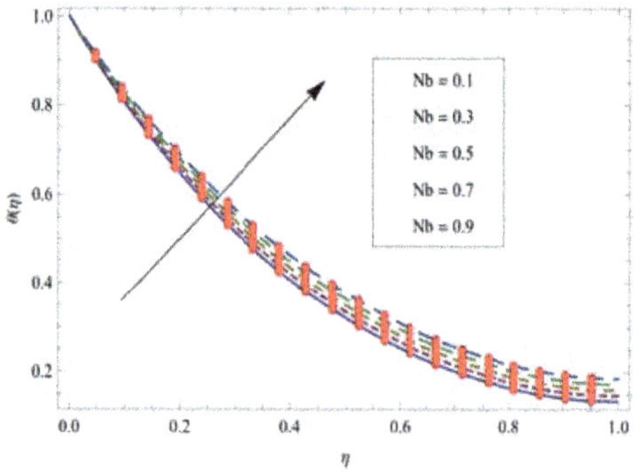

Figure 19. Illustrates the effect of Nb on $\theta(\eta)$, when $h = -0.5$, $\lambda = 0.7, S = 0.7, \mathrm{kr} = 0.7, \mathrm{Pr} = 30$, $\mathrm{Ec} = 0.7, \mathrm{Nt} = 0.5, \beta = 1$ and $Sc = 0.7$.

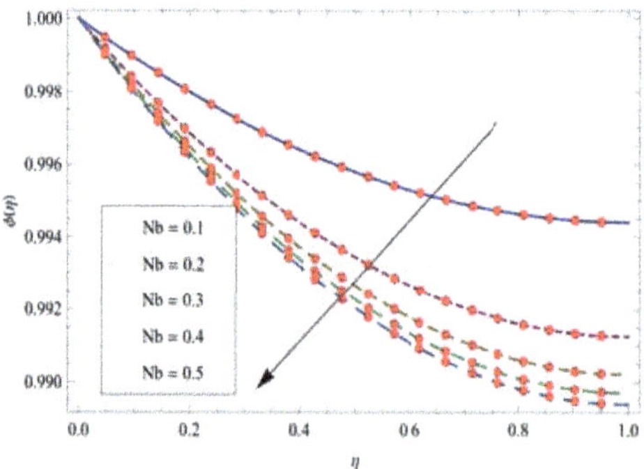

Figure 20. Indicates the effect of Nb on $\phi(\eta)$, for $h = -0.9$, $\lambda = 0.5, S = 0.7, \mathrm{kr} = 0.5, \mathrm{Pr} = 0.5$, $\mathrm{Ec} = 0.5, \mathrm{Nt} = 0.7, \beta = 1$ and $Sc = 0.5$.

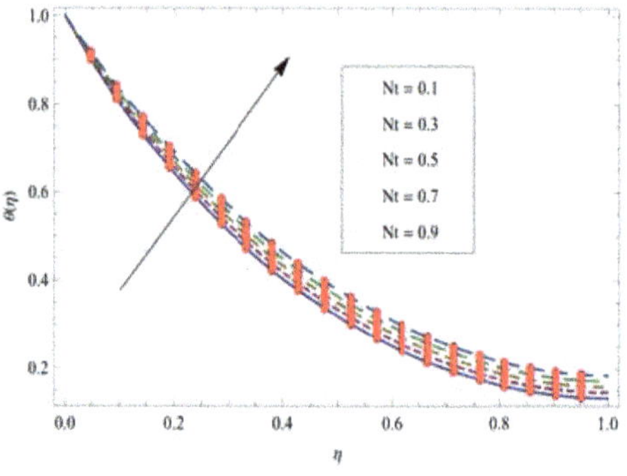

Figure 21. The effect of Nt on $\theta(\eta)$, when $h = -0.5$, $\lambda = 0.7, S = 0.7, \mathrm{kr} = 0.7, \mathrm{Pr} = 30, \mathrm{Ec} = 0.7$, $\mathrm{Nb} = 0.5, \beta = 1$, and $Sc = 0.7$.

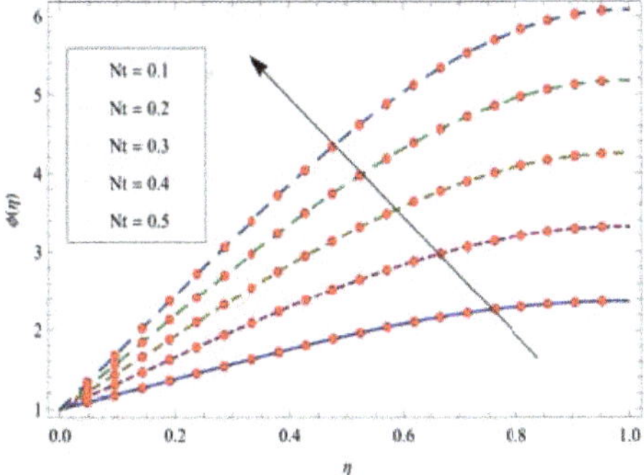

Figure 22. The effect of Nt on $\phi(\eta)$, for $h = -0.9$, $\lambda = 0.5, S = 0.7, \text{kr} = 0.5, \text{Pr} = 0.5, \text{Ec} = 0.5$, $\text{Nb} = 0.7, \beta = 1$ and $Sc = 0.5$.

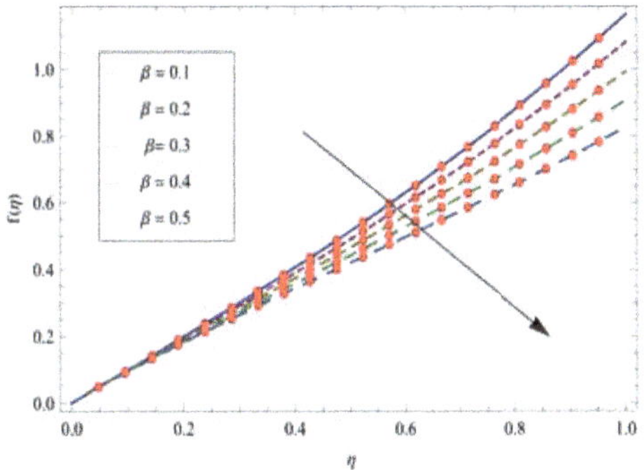

Figure 23. Shows the effect of β on $f(\eta)$, for $h = -0.7, \lambda = 1, \text{kr} = 0.5, \text{Pr} = 0.5, \text{Ec} = 0.5$, $\text{Nb} = 0.5$, $\text{Nt} = 0.6, \beta = 0.1$ and $Sc = 0.6$.

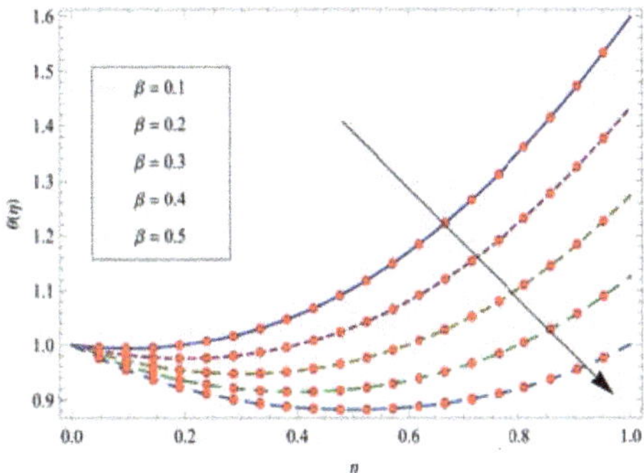

Figure 24. The effect of β on $\theta(\eta)$, for $h = -0.7$, $\lambda = 1, S = 0.5, \text{kr} = 0.5, \text{Ec} = 0.5, \text{Nb} = 0.5, \text{Nt} = 0.6$ and $Sc = 0.6$.

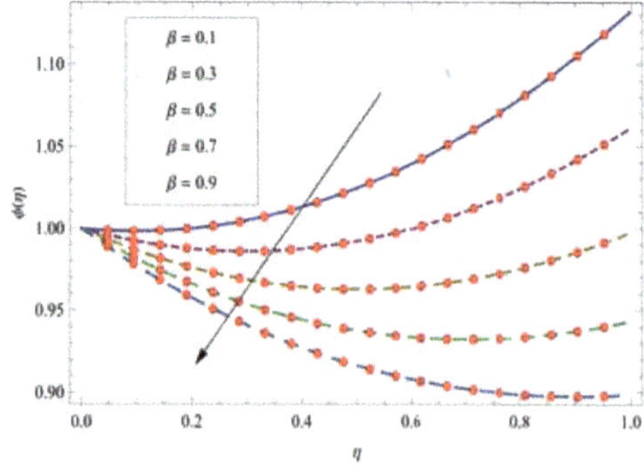

Figure 25. Illustrates the effect of β on $\phi(\eta)$ for $h = -0.25$, $\lambda = 0.5, S = 0.1, \mathrm{kr} = 0.5, \mathrm{Ec} = 0.5$, $\mathrm{Nb} = 0.5, \mathrm{Nt} = 0.6$ and $Sc = 0.6$.

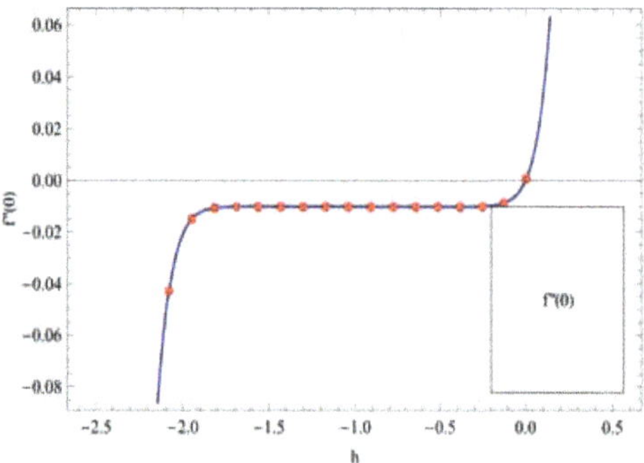

Figure 26. Depicts h curves of $f''(0)$, when $\lambda = 0.9, \mathrm{kr} = 0.9, \mathrm{Pr} = 0.5, \mathrm{Ec} = 0.5, \mathrm{Nb} = 0.5$, $\mathrm{Nt} = 0.6$, $\beta = 1$ and $Sc = 0.6$.

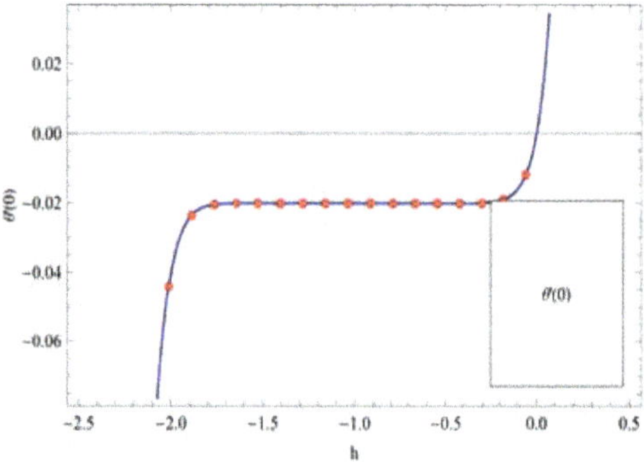

Figure 27. Shows h curves of $\theta'(0)$, for $\lambda = 0.2, S = 0.2, \mathrm{kr} = 0.2, \mathrm{Pr} = 1, \mathrm{Ec} = 0.6, \mathrm{Nb} = 1, \mathrm{Nt} = 0.1$, $\beta = 1$ and $Sc = 0.5$.

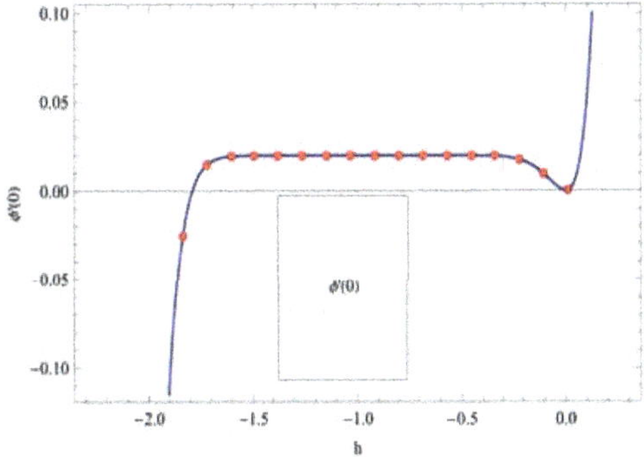

Figure 28. Elaborates h curves of $\phi'(0)$, when $\lambda = 0.2, S = 0.2, \text{kr} = 0.2, \text{Pr} = 1, \text{Ec} = 0.6, \text{Nb} = 0.4, \text{Nt} = 0.5, \beta = 1, Sc = 0.5$.

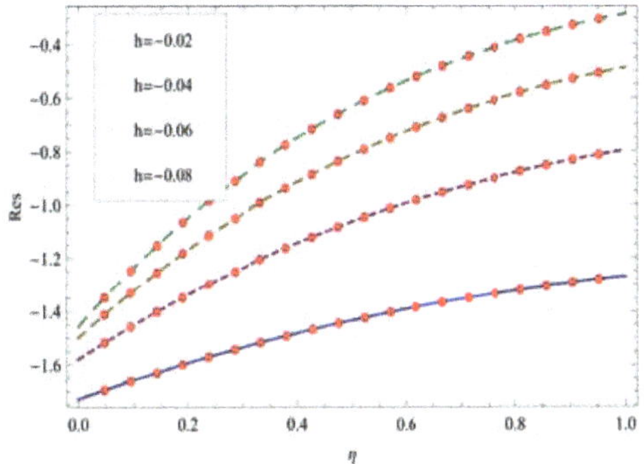

Figure 29. Illustrates h curves of the residuals for the velocity profile $f(\eta)$, when λ = 0.6, S = 0.6, Kr = 0.4, Pr = 1, Ec = 0.4, Nb = 0.6, Nt = 0.5, Sc = 0.5, $\beta = 1$.

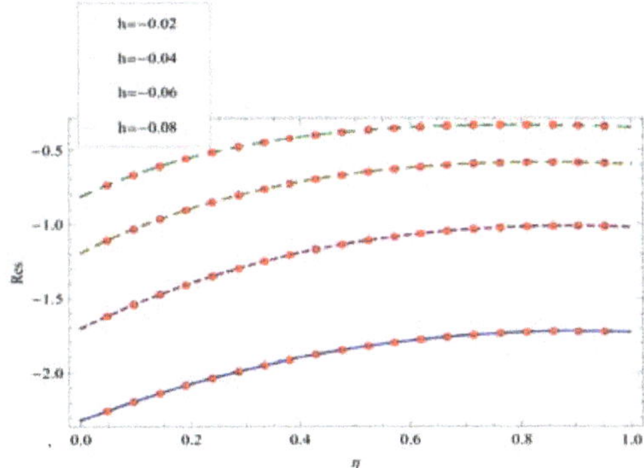

Figure 30. Indicates h curves of the residuals for the temperature profile $\theta(\eta)$, when $\lambda = 0.6$, $S = 0.6$, $\text{Kr} = 0.4, \text{Pr} = 1, \text{Ec} = 0.4, \text{Nb} = 0.6, \text{Nt} = 0.5, \text{Sc} = 0.5, \beta = 1$.

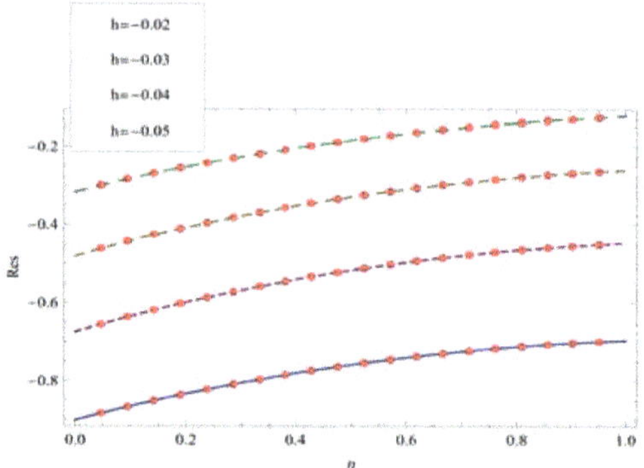

Figure 31. Shows h curves of the residuals for the concentration profile $\phi(\eta)$, when $\lambda = 0.6, s = 0.6$; $\mathrm{Kr} = 0.4, \mathrm{Pr} = 1, \mathrm{Ec} = 0.4, \mathrm{Nb} = 0.6, \mathrm{Nt} = 0.5, \mathrm{sc} = 0.5, \beta = 1$.

Table 1. Shows the Nomenclature.

Alphabet	Defined as	Alphabet	Defined as
x	horizontal coordinate (m)	T_r	Reference temperature
y	vertical coordinate (m)	T_0	initial temperature of the fluid (K)
u	horizontal velocity component (m/s)	T	temperature (K)
v	vertical velocity component (m/s)	U_w	Velocity of the stretching sheet
S	Unsteadiness parameter	T_1	final temperature of the fluid (K)
T_w	Temperature at the sheet	T	temperature (K)
K	thermal diffusivity (m^2)	t	time (s)
S	Unsteadiness parameter	f	Dimensionless Velocity
b	stretching parameter(constant)	k'	permeability coefficient of the porosity
$f(\xi)$	nondimensional variable for velocity	C_w	Nanoparticle volume fraction at sheet
Pr	Prandtl number	Ec	Eckert number
Kr	nondimensional porosity parameter	Sc	Schmidt number
Nb	Brownian motion parameter	D_B	Brownian diffusion coefficient
C_p	specific heat at constant pressure (kJ kg^{-1} K^{-1})	Nt	Thermophoresis parameter
Greek symbols	**Defined as**	**Greek Symbols**	**Defined as**
ϕ	Dimensionless nanoparticle volume fraction	ξ	Similarity variable
ν	kinematic viscosity of the fluid	ρ_f	Density of base fluid
ρ	density (kg m^{-3})	$(\rho c)_p$	Heat capacity of the nanoparticle material
Γ	Time constant	α	Thermal diffusivity of the base fluid
ρ_p	Nanoparticle mass density	$(\rho c)_f$	Heat capacity of the base fluid
ν	Kinematic viscosity of the base fluid	λ	Williamson fluid constant
β	non-dimensional film thickness	θ	Dimensionless temperature
ψ	non-dimensional stream function	$(')$	differentiation w. r. t. ξ

Table 2. HAM, Numerical Solution and their absolute Error are shown for $f(\eta)$, when $h = -0.47$, $\lambda = 0.2, \mathrm{kr} = 0.2, \mathrm{Pr} = 1, \mathrm{Ec} = 0.6, \mathrm{Nb} = 0.4, \mathrm{Nt} = 0.5, S = 0.2, \beta = 1$ and $Sc = 0.5$.

η	Numerical Solution for $f(\eta)$	HAM Solution for $f(\eta)$	Absolute Error
0	0	0	0
0.1	0.0952761	0.0955184	2.4×10^{-4}
0.2	0.182152	0.182958	8.1×10^{-4}
0.3	0.262045	0.263561	1.5×10^{-3}
0.4	0.336198	0.338462	2.3×10^{-3}
0.5	0.405709	0.408698	2.9×10^{-3}
0.6	0.471556	0.47522	3.6×10^{-3}
0.7	0.534618	0.5389	4.3×10^{-3}
0.8	0.595686	0.600535	4.6×10^{-3}
0.9	0.655482	0.660862	5.4×10^{-3}
1	0.714666	0.720558	5.8×10^{-3}

Table 3. HAM, Numerical Solution and their absolute Error are elobarated for $\theta(\eta)$, when $h = -0.5$, $\lambda = 0.2, \mathrm{kr} = 0.2, \mathrm{Pr} = 0.5, \mathrm{Ec} = 0.6, \mathrm{Nb} = 0.4, \mathrm{Nt} = 0.5, S = 0.2, \beta = 1$ and $Sc = 0.7$.

η	Numerical Solution for $\theta(\eta)$	HAM Solution for $\theta(\eta)$	Absolute Error
0	1	1	1.4×10^{-8}
0.1	0.907342	0.90737	2.8×10^{-5}
0.2	0.82775	0.827596	1.5×10^{-4}
0.3	0.759861	0.759482	3.8×10^{-4}
0.4	0.702621	0.702067	5.5×10^{-4}
0.5	0.655233	0.654586	6.5×10^{-4}
0.6	0.617098	0.616438	6.7×10^{-4}
0.7	0.587786	0.587169	6.2×10^{-4}
0.8	0.567002	0.566451	5.5×10^{-4}
0.9	0.554573	0.554078	4.9×10^{-4}
1	0.550428	0.549956	4.7×10^{-4}

Table 4. HAM, Numerical Solution and their absolute Error are depicted for $\phi(\eta)$, when $h = -0.6$, $\lambda = 0.2, \mathrm{kr} = 0.2, \mathrm{Pr} = 1, \mathrm{Ec} = 0.6, \mathrm{Nb} = 1, \mathrm{Nt} = 0.1, S = 0.2, \beta = 1$, and $Sc = 0.5$.

η	Numerical Solution for $\phi(\eta)$	HAM Solution for $\phi(\eta)$	Absolute Error
0	1	1	2.9×10^{-9}
0.1	0.940581	0.941167	5.9×10^{-4}
0.2	0.890316	0.8912	8.8×10^{-4}
0.3	0.848224	0.849209	9.8×10^{-4}
0.4	0.813465	0.81443	9.6×10^{-4}
0.5	0.785325	0.786207	8.8×10^{-4}
0.6	0.763201	0.76398	7.8×10^{-4}
0.7	0.74659	0.747277	6.9×10^{-4}
0.8	0.735082	0.735705	6.2×10^{-4}
0.9	0.728352	0.728942	5.9×10^{-4}
1	0.726151	0.726734	5.8×10^{-4}

Table 5. Illustrates the residuals achieved by HAM for system of coupled differential equations forming in velocity, temperature and concentration profiles.

η	Residuals for $f(\eta)$	Residuals for $\theta(\eta)$	Residuals for $\phi(\eta)$
0	-2.0×10^{-1}	-3.7×10^{-1}	5.1×10^{-2}
0.1	-1.2×10^{-1}	-7.9×10^{-2}	2.4×10^{-2}
0.2	-4.9×10^{-2}	3.2×10^{-3}	2.2×10^{-2}
0.3	-4.9×10^{-4}	2.5×10^{-2}	2.9×10^{-2}
0.4	3.3×10^{-2}	2.9×10^{-2}	3.2×10^{-2}
0.5	5.2×10^{-2}	2.1×10^{-2}	2.3×10^{-2}
0.6	6.1×10^{-2}	5.9×10^{-3}	2.2×10^{-3}
0.7	5.9×10^{-2}	-1.3×10^{-2}	-2.6×10^{-2}
0.8	4.9×10^{-2}	-3.3×10^{-2}	-5.5×10^{-2}
0.9	3.3×10^{-2}	-4.6×10^{-2}	-7.3×10^{-2}
1	9.9×10^{-3}	-4.9×10^{-2}	-8.7×10^{-2}

6. Conclusions

The main conclusion of this endeavor is the study of liquid film in a porous medium considering non-Newtonian Williamson fluid on an unstable stretching surface. The effect of Thermophoresis and Brownian motion has been countered to the liquid film flow. The solutions of the problems have

been achieved by using analytical technique, HAM for velocity, temperature and concentration fields respectively. The influences of all parameters included in the problem have been described and the solutions are displayed in the diagrams for checking their effects on velocity, temperature as well as concentration fields. The coupled problem has been solved by using an analytical method HAM. The h curves for the residuals of velocity, temperature and concentration have been sketched.
The main concluded points are derived as,

(1) Increasing thickness parameter β produce the friction force and as a result velocity of the fluid film falls down.
(2) The larger values of β transport more fluid in the boundary layer region and cooling effect is produced which absorbed the heat transfer from the sheet and as a result the temperature reduces.
(3) The Eckert number Ec is allied with the viscous dissipation term and lead to incrrease the quantity of heat being produced by the shear forces in the fluid. Therefore, larger values of Ec raises the temperature field.
(4) The larger values of Prandtl number Pr reduces the thermal boundary layer due to which the temperature field reduces.
(5) Higher values of Porosity parameter Kr generate larger open space and create hurdle to flow and as a result the flow field reduces.

Author Contributions: Liaqat Ali, Taza Gul and Waris Khan modeled the problem and solved it; Saeed Islam, Liaqat Ali, and Taza Gul contributed to the discussion of the problem; L.C.C.Dennis, Saeed Islam, Ilyas Khan and Aurangzeb Khan contributed in the English corrections. All the authors read and approved the final manuscript.

Conflicts of Interest: The authors declare no conflict of interest.

References

1. Williamson, R.W. The dyanamics of lava flows. *Annu. Rev. Fluid Mech.* **2000**, *32*, 477–518.
2. Dapra, I.; Scarpi, G. Perturbation solution for pulsatile flow of a non newtonian Williamson fluids in rock fracture. *Int. J. Rock Mech. Min. Sci.* **2006**, *44*, 1–8.
3. Wang, C.Y. Liquid film on an unsteady stretching surface. *Q. Appl. Math.* **1990**, *48*, 601–610.
4. Cramer, K.; Pai, S. *Magneto Fluid Dynamics for Engineers and Applied Physicists*; McGraw-Hill: New York, NY, USA, 1973.
5. Selim, A.; Hossain, M.A.; Rees, D.A.S. The effect of surface mass transfer mixed convection flow past a heated vertical flat permeable plate with thermophoresis. *Int. J. Therm. Sci* **2003**, *42*, 973–982.
6. Das, K. Impact of thermal radiation on MHD slip flow over a flat plate with variable fluid properties. *Heat Mass. Transf.* **2012**, *48*, 767–778.
7. Siddeshwar, P.G; Mahabaleshwar, U.S. Effects of radiation and heat transfer on MHD flow of viscoelastic liquid and heat transfer over a stretching sheet. *Int. J. Nonlinear Mech.* **2005**, *40*, 807–820.
8. Nadeem, S.; Hussain, S.T. Flow and heat transfer analysis of Williamson nanofluid. *Appl. Nanosci.* **2014**, *4*, 1005–1012.
9. Hassanein, I.A.; Essawy, A.; Morsy, N.M. Variable viscosity and thermal conductivity effects on heat transfer by natural convection from a cone and a wedge in porous media. *Arch. Mech.* **2004**, *55*, 345–356.
10. Aziz, R.C.; Hashim, I.; Alomari, A.K. Thin film flow and heat transfer on an unsteady stretching sheet with internal heating. *Meccanica* **2011**, *46*, 349–357.
11. Qasim, M.; Khan, Z.H.; Lopez, R.J.; Khan, W.A. Heat and mass transfer in nanofluid over an unsteady stretching sheet using Buongiorno's model. *Eur. Phys. J. Plus* **2016**, *131*, 1–16.
12. Mahesh, K.; Gireesha, B.J.; Rama, S.R.G. Heat and Mass Transfer in Nanofuid over an unsteady stretching surface. *J. Nanofluids* **2015**, *4*, 1–8.
13. Ellahi, R.; Hassan, M.; Zeeshan, A. Aggregation effects on water base Al_2O_3-Nanofluid over permeable wedge in mixed convection. *Asia-Pac. J. Chem. Eng.* **2016**, *11*, 179–186.
14. Akbar, N.S.; Raza, M.; Ellahi, R. CNT suspended CuO + H_2O nano fluid and energy analysis for the peristaltic flow in a permeable channel. *Eng. J.* **2015**, *54*, 623–633.
15. Akbar, N.S.; Raza, M.; Ellahi, R. Copper oxide nanoparticles analysis with water as base fluid for peristaltic flow in permeable tube with heat transfer. *Comput. Methods Progr. Biomed.* **2016**, *130*, 22–30.

16. Shehzad, N.; Zeeshan, A.; Ellahi, R.; Vafai, K. Convective heat transfer of nanofluid in a wavy channel: Buongiorno's mathematical model. *J. Mol. Liq.* **2016**, *222*, 446–455.
17. Zeeshan, A.; Hassan, M.; Ellahi, R.; Nawaz, M. Shape effect of nanosize particles in unsteady mixed convection flow of nanofluid over disk with entropy generation. *J. Process Mech. Eng.* **2016**, doi:10.1177/0954408916646139.
18. Khan, W.; Gul, T.; Idrees, M.; Islam, I.; Khan, I.; Dennis, L.C.C. Thin Film Williamson Nanofluid Flow with Varying Viscosity and Thermal Conductivity on a Time-Dependent Stretching Sheet. *Appl. Sci.* **2016**, *6*, 334, doi:10.3390/app6110334.
19. Gamal, M.; Rahman, A. Effect of Magnetohydrodynamic on Thin Films of Unsteady Micropolar Fluid through a Porous Medium. *J. Mod. Phys.* **2011**, *2*, 1290–1304 .
20. Jaina, S.; Choudhary, R. Effects of MHD on Boundary Layer Flow in Porous Medium due to Exponentially Shrinking Sheet with Slip. *Procedia Eng.* **2015** , *127*, 1203–1210.
21. Liao, S. *Beyond Perturbation: Introduction to the Homotopy Analysis Method;* Chapman and Hall/CRC: Boca Raton, FL, USA, 2003.
22. Liao, S.J. An optimal homotopy-analysis approach for strongly nonlinear differential equations. *Commun. Nonlinear Sci. Numer. Simul.* **2010**, *15*, 2003–2016.
23. Liao, S. On the homotopy analysis method for nonlinear problems. *Appl. Math. Comput.* **2004**, *147*, 499–513.
24. Abbasbandy, S.; Shirzadi, A. A new application of the homotopy analysis method: Solving the Sturm—Liouville problems. *Commun. Nonlinear Sci. Numer. Simul.* **2011**, *16*, 112–126.
25. Abbasbandy, S. Homotopy analysis method for heat radiation equations. *Int. Commun. Heat Mass Transf.* **2007**, *34*, 380–388.
26. Abbasbandy, S. The application of homotopy analysis method to solve a generalized Hirota-Satsuma coupled KdV equation. *Phys. Lett. A* **2007**, *361*, 478–483.
27. Khan, N.S.; Taza Gul, T.; Islam, S.; Khan, W. Thermophoresis and thermal radiation with heat and mass transfer in a magnetohydrodynamic thin film second grade fluid of variable properties past a stretching sheet. *Eur. Phys. J. Plus* **2017**, *132*, doi:10.1140/epjp/i2017-11277-3.
28. Khan, W.; Gul, T.; Idrees, M.; Khan, I. Dufour and Soret Effect with Thermal Radiation on the Nano Film Flow of Williamson Fluid Past Over an Unsteady Stretching Sheet. *J. Nanofluids* **2017**, *6*, 243–253.
29. Narayana, M.; Sibanda, P. Hydromagnetic nanofluid flow due to a stretching or shrinking sheet with viscous dissipation and chemical reaction effects. *Int. J. Heat Mass. Transf.* **2012**, *55*, 7587–7595.
30. Xu, H.; Pop, I.; You, X.C. Flow and heat transfer in a nano-liquid film over an unsteady stretch surface. *Int. J. Heat Mass. Transf.* **2013**, *60*, 646–652.
31. Nadeem, S.; Haq, R.U.; Lee, C. MHD flow of a Casson fluid over an exponentially shrinking sheet. *Sci. Iran.* **2012**, *19*, 1550–1553.
32. Nadeem, S.; Haq, R.U.; Akbar, N.S.; Khan, Z.H. MHD three-dimensional Casson fluid flow past a porous linearly stretching sheet. *Alex. Eng. J.* **2013**, *52*, 577–582.
33. Rehman, S.U.; Haq, R.U.; Lee, C.; Nadeem, S. Numerical study of non-Newtonian fluid flow over an exponentially stretching surface: an optimal HAM validation. *Braz. Soc. Mech. Sci. Eng.* **2016**, *7*, doi:10.1007/s40430-016-0687-3.
34. Haq, R.U.; Shahzad, F.; Al-Mdallal, Q.M. MHD Pulsatile Flow of Engine oil based Carbon Nanotubes between Two Concentric Cylinders. *Results Phys.* **2016**, *7*, 57–68.
35. Besthapu, P.; Haq, R.U.; Bandari, S.; Al-Mdallal, Q.M. Mixed convection flow of thermally stratified MHD nanofluid over an exponentially stretching surface with viscous dissipation effect. *J. Taiwan Inst. Chem. Eng.* **2017**, *71*, 307–314.

Article

Viscosity Prediction of Different Ethylene Glycol/Water Based Nanofluids Using a RBF Neural Network

Ningbo Zhao and Zhiming Li *

College of Power and Energy Engineering, Harbin Engineering University, Harbin 150001, China; zhaoningbo314@126.com

* Correspondence: lizhimingheu@126.com; Tel.: +86-451-8251-9647

Academic Editor: Rahmat Ellahi

Received: 15 March 2017; Accepted: 14 April 2017; Published: 18 April 2017

Abstract: In this study, a radial basis function (RBF) neural network with three-layer feed forward architecture was developed to effectively predict the viscosity ratio of different ethylene glycol/water based nanofluids. A total of 216 experimental data involving CuO, TiO_2, SiO_2, and SiC nanoparticles were collected from the published literature to train and test the RBF neural network. The parameters including temperature, nanoparticle properties (size, volume fraction, and density), and viscosity of the base fluid were selected as the input variables of the RBF neural network. The investigations demonstrated that the viscosity ratio predicted by the RBF neural network agreed well with the experimental data. The root mean squared error (*RMSE*), mean absolute percentage error (*MAPE*), sum of squared error (*SSE*), and statistical coefficient of multiple determination (R^2) were respectively 0.04615, 2.12738%, 0.46007, and 0.99925 for the total samples when the Spread was 0.3. In addition, the RBF neural network had a better ability for predicting the viscosity ratio of nanofluids than the typical Batchelor model and Chen model, and the prediction performance of RBF neural networks were affected by the size of the data set.

Keywords: nanofluids; viscosity; RBF neural network; ethylene glycol/water

1. Introduction

As a very important heat transfer medium, ethylene glycol/water mixtures are widely used in many different kinds of industrial equipment including car radiators, air conditioning systems, and liquid cooled computers [1]. In the past few decades, with the rapid development of various compact heat exchange components, the conventional ethylene glycol/water mixtures have been unable to effectively meet the ever-increasing demand for cooling due to their lower thermal conductivity. Therefore, how to develop enhanced heat transfer technology has become a very important problem in the fields of thermal engineering [2].

Nanofluids, a special liquid-solid mixture containing a base fluid and nanoparticles (usually less than 100 nm), have drawn increasing attention recently because of their advantages in thermal conductivity and stability [3]. Many investigations indicated that nanofluids could be an effective technology to improve the heat transfer performance of systems using ethylene glycol/water mixtures as coolant [4]. For example, the experimental results of Vajjha and Das [5] showed that at the temperature of 299 K, the thermal conductivities of the 60:40 (by weight ratio) ethylene glycol/water mixture could be increased by about 12.3% by adding ZnO nanoparticles (29 nm) with a volume fraction of 2%. Sundar et al. [6] experimentally investigated the effects of Fe_3O_4 nanoparticles (13 nm) on three different kinds of ethylene glycol/water mixtures with weight ratios of 20:80, 40:60, and 60:40. They found that at the temperature of 60 °C and the nanoparticle volume fraction of 2%, the thermal conductivity enhancements of the above three ethylene glycol/water mixtures were 46%, 42%, and 33%, respectively.

Thermo-physical parameters are very important factors that affect the heat and mass transfer performance of nanofluids [7–9]. Due to the fact that viscosity can significantly affect the flow internal resistance, inlet Reynolds number, and pressure drop, many experimental investigations have been carried out regarding the viscosity of different nanofluids. As reported by Azmi et al. [10], the viscosity of the 40:60 (by volume ratio) ethylene glycol/water mixture could be increased obviously by dispersing TiO_2 nanoparticles. For example, the viscosity enhancement was about 12% when the nanoparticle volume fraction changed from 0.5% to 1.5%. Sundar et al. [11] investigated the viscosity variations of Fe_3O_4-ethylene glycol/water nanofluids with different nanoparticle fractions and working temperatures. Their experimental results indicated that the viscosity of ethylene glycol/water based nanofluids could be increased by increasing the nanoparticle volume fraction and decreasing temperature. At a nanoparticle volume fraction of 1%, the viscosity of the base fluid could be enhanced by 2.9 times. Chen et al. [12], Jamshidi et al. [13], Kulkarni et al. [14], Rudyak et al. [15], Namburu et al. [16], Lim et al. [17], Chiam et al. [18], and Li et al. [19] respectively measured the viscosity of various ethylene glycol/water mixture based nanofluids with the effects of different factors. According to their experimental results, it was found that a suspension of nanoparticles could enhance the viscosity of the base fluid in different degrees. Additionally, temperature, base fluid, and nanoparticle properties including volume fraction, size, type, and shape were the important factors affecting the enhancement of nanofluids' viscosity.

For the basis of the experimental research, the modeling and prediction of viscosity is also very important for understanding the rheological behavior of nanofluids. Murshed and Estellé [20] reviewed the latest developments of viscosity models for nanofluids. Their analysis indicated that although many theoretical models and empirical correlations have been developed for nanofluid viscosity, only a few of them were used for ethylene glycol/water based nanofluids. Additionally, since the effects of different factors on nanofluid viscosity were usually coupled and uncertain, it was still very difficult to accurately describe the viscosity characteristics of different nanofluids in a wide range of nanoparticle volume fractions, sizes, temperatures, etc. Therefore, how to develop an effective solution for the viscosity prediction of nanofluids is a hot topic in the field of nanofluids.

Artificial neural networks (ANN), a black box data analysis approach, has a strong nonlinear mapping ability to establish the relationship between input and output variables without considering the detailed physical process. Due to the advantages of ANNs such as high speed, simplicity, and large capacity, various ANNs were put forward to solve the modeling and prediction problems of nanofluid viscosity [21]. Selecting five variables (temperature, nanoparticle volume fraction, nanoparticle size, viscosity of the base fluid, and relative density of the base fluid) and nanoparticles as the input, Yousefi et al. [22] developed a diffusional neural network (DNN) to predict the viscosity of six different types of nanofluids. As reported in their analysis, DNN could be used for predicting the viscosity of nanofluids with satisfactory accuracy. On this basic, Mehrabi et al. [23] analyzed the application of a Fuzzy C-Means-based Adaptive neuro-fuzzy inference system (FCM-ANFIS) for the viscosity prediction of various water based nanofluids. They found that the FCM-ANFIS predicted values agreed well with the experimental data. Attracted by the better nonlinear mapping and recognition abilities of ANN, Zhao et al. [24,25] investigated the feasibility of RBF neural networks for predicting the viscosity of two water based nanofluids containing Al_2O_3 and CuO nanoparticles. Their results demonstrated that ANN was an effective tool in comparison with the traditional model-based approach for describing the enhancement behavior of nanofluid viscosity. They indicated that the addition of temperature as an input variable could improve the prediction performance of the RBF neural network.

To the best of the authors' knowledge, there are few publications that study the modeling and prediction of different ethylene glycol/water based nanofluids using ANN. Considering the advantages of RBF neural networks that are easier to design, and have faster training speed, higher training accuracy, stronger generalization ability, and stronger tolerance for input noise [26], this paper selects a RBF neural network as a competitive method for predicting the viscosity characteristics of different ethylene glycol/water based nanofluids with different influence factors. Firstly, the basis theory and modeling process of the RBF neural network are introduced briefly. On this basis, the available measurements from various published studies are obtained to establish the data sample sets and train the RBF neural network for determining the network configuration. Finally, the RBF neural networks' predicted results are compared with the experimental data to evaluate the prediction performance of the proposed model.

2. Basic Theory of a RBF Neural Network

Benefiting from the inspiration of the human brain's structure and activity mechanism, many different artificial neural networks have been developed for different purposes including classification and regression. In the fields of curve-fitting and nonlinear predictive modeling, the RBF neural network proposed by Broomhead and Lowe [27] can exhibit a good ability because of its high accuracy and stability [28].

Figure 1 presents the basic structure of a typical three-layer RBF neural network. The input and output layers respectively correspond to the dendrite and synapse of biological neurons, which are used to mathematically describe the modeling object. The hidden layer, similar to the function of the cyton, plays a role of intermediation to process the input-output information and deliver it to the output layer. The connections between different layers are established through a series of artificial neurons and weights.

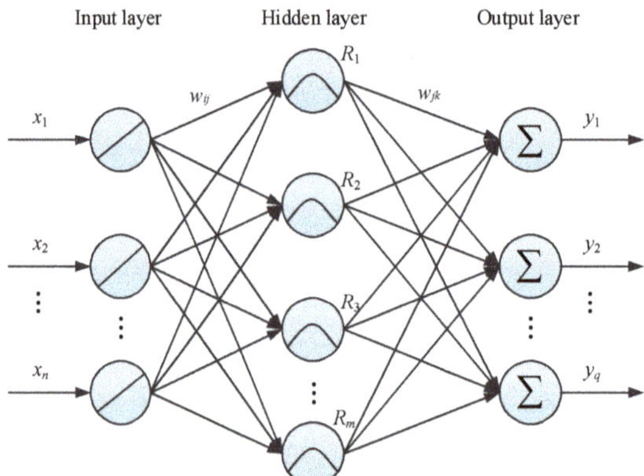

Figure 1. A typical three-layer RBF neural network.

Theoretically, the modeling process of the RBF neural network is to solve the mapping from X^n to Y^q $(n, q \geq 1)$ in Euclidean space. Assuming that the input vector of the RBF neural network is X, the response of the kth neuron in the output layer ($y_k \in Y^q$) can be obtained by using the following linear weighting function [29].

$$y_k = \sum_{j=1}^{m} \omega_{jk} R_j(X), \quad (k = 1, 2, \cdots, q) \tag{1}$$

where ω_{jk} is the connection weight between the jth hidden layer neuron and the kth output layer neuron. m and q are the numbers of neurons in the corresponding layer, respectively.

Different from many other ANNs, the response of the RBF neural network's jth hidden layer neuron is usually determined by the RBF. When it selects a Gaussian function, the corresponding $R_j(X)$ can be defined as,

$$R_j(X) = \exp(-\frac{\|X-c_j\|^2}{2\sigma_j^2}), \quad (j = 1, 2, \cdots, m) \tag{2}$$

where $\|\|$ is the Euclidean distance between the input vector X and the jth neuron center c_j. σ_j is the width of the jth neuron.

Analyzing Equations (1) and (2), it can be easily found that the key of RBF neural network training is how to determine ω_{jk}, c_j, and σ_j. In the past few decades, different unsupervised and supervised algorithms have been developed to solve this problem [30]. In this study, the network parameters are updated by using an orthogonal least squares (OLS) approach, of which the minimizing function is shown in Equation (3). More detailed information about OLS can be found in [31].

$$\min J = \sum_{k=1}^{q} (|y_{nk} - y_{dk}|^2) \tag{3}$$

where y_{nk} and y_{dk} are the network output and desired output of the kth output layer node, respectively.

3. Modeling Implementation of a RBF Neural Network

According to the above theory, the modeling process of the RBF neural network involves three main parts which are data preparation, training, and testing. Figure 2 depicts the basic applied flow chart of the RBF neural network for predicting the relative viscosity of ethylene glycol/water based nanofluids. The specific implementations are discussed in the following.

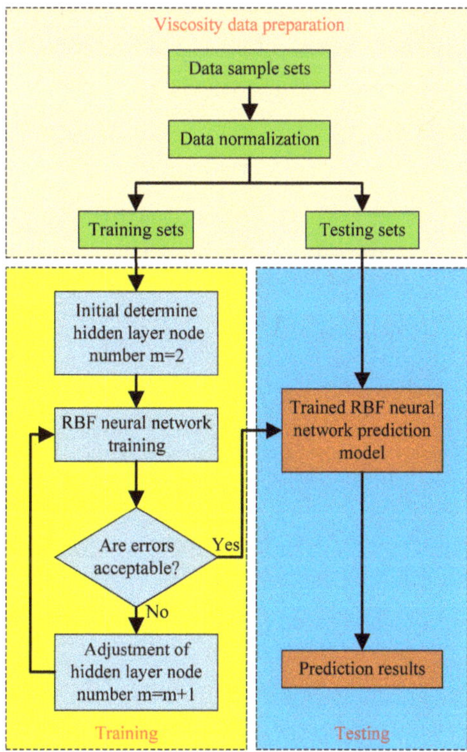

Figure 2. Implementation process of a RBF neural network for viscosity prediction.

3.1. Preparation of Viscosity Data

As previously mentioned, many experimental investigations have been published to discuss the effects of different factors including temperature and nanoparticle properties (such as type, size, concentration, and shape) on the viscosity of ethylene glycol/water based nanofluids. Considering the integrity of the measuring information, a total of 216 viscosity data involving TiO_2, CuO, SiO_2, and SiC are obtained to establish the sample sets. The detailed information of the nanofluids regarding nanoparticle diameter (d_p), nanoparticle volume fraction (ϕ_p), nanoparticle density (ρ_p), temperature (T), and viscosity of the base fluid (μ_f) and nanofluids (μ_{nf}) are listed in Table 1. According to the modeling principle, 198 data (about 90%) are selected to train the RBF neural network, and the remaining 18 data (about 10%) are used to test the performance of the trained RBF neural network.

Table 1. Viscosity information of ethylene glycol/water based nanofluids.

Nanofluids	TiO_2-EG [12]	SiO_2-EG/W [a] [13]	SiO_2-EG/W [b] [13]	CuO-EG/W [c] [14]	SiO_2-EG [15]	CuO-EG/W [c] [16]	SiC-EG/W [d] [19]
d_p (nm)	25	10	10	30/45/50	18.1/28.3/45.6	29	30
ϕ_p (%)	0.1–1.8	0.1	0.1	1–6	0.6–8.4	1–4	0.1–0.5
ρ_p (kg/m^3)	4230	2650	2650	6310	2650	6310	110
T (°C)	20.1–60.2	28.45–59	28–59	−35–50	25–59	0–40	10–50
μ_f (mPa.s)	3.87–23	0.98–1.68	1.6–3.11	2.33–99.5	4.08–18.5	4.35–11.5	9.2–11.34
μ_{nf}/μ_f	0.81–1	1–1.15	1.05–1.13	1.1–4.65	1.04–2.02	1.14–2.09	1.13–1.29
No. of data	27	11	10	80	31	12	45

[a] EG/W: 25:75 by volume ratio; [b] EG/W: 50:50 by volume ratio; [c] EG/W: 60:40 by weight ratio; [d] EG/W: 40:60 by weight ratio.

To improve the learning and training performances of the RBF neural network, the following equation is used to normalize the input and output variables.

$$x' = \frac{x - x_{\min}}{x_{\max} - x_{\min}} \tag{4}$$

where x is the original value, x' is the normalized value, and $x_{\max}$ and $x_{\min}$ are the corresponding maximum and minimum of x.

3.2. Configuration of a RBF Neural Network

Considering the nonlinear characteristics of the ethylene glycol/water based nanofluid viscosity ratio with different factors, a three layer RBF neural network is developed in the present investigation. Temperature, nanoparticle diameter, nanoparticle volume fraction, nanoparticle density, and viscosity of the base fluid are selected as the input variables. The objective output of the RBF neural network is the viscosity ratio between the nanofluids and the base fluid. Therefore, the basic structure of the developed RBF neural network for predicting the viscosity ratio of ethylene glycol/water based nanofluids is 5-m-1, as illustrated in Figure 3. For the neurons, the numbers in the hidden layer (m) and other parameters are determined in the training process.

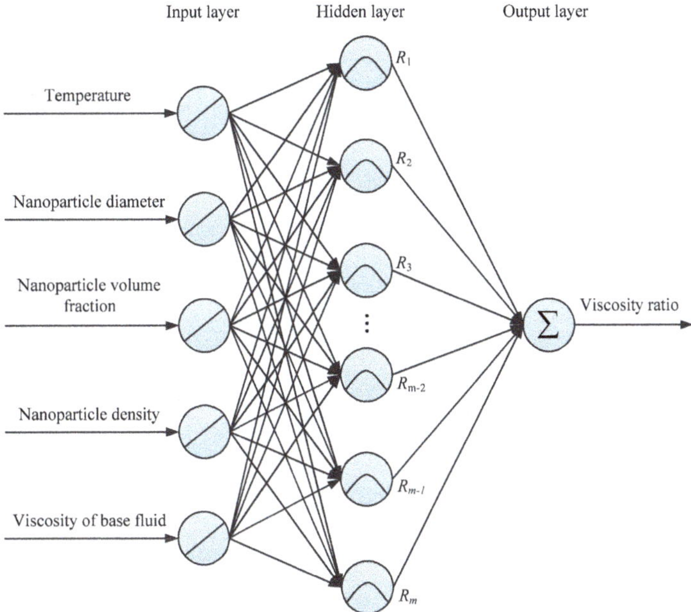

Figure 3. RBF neural network developed in this study.

3.3. Evaluation Criteria

To effectively evaluate the training and prediction performance of the RBF neural network, the following four important parameters are used.

Root mean squared error (*RMSE*),

$$RMSE = \left(\frac{1}{t}\sum_{l=1}^{t}|P_l - Q_l|^2\right)^{1/2} \tag{5}$$

Mean absolute percentage error (*MAPE*),

$$MAPE = \frac{100\%}{t}\sum_{l=1}^{t}\left|\frac{P_l - Q_l}{P_l}\right| \tag{6}$$

Sum of squared error (*SSE*),

$$SSE = \sum_{l=1}^{t}(P_l - Q_l)^2 \tag{7}$$

Statistical coefficient of multiple determination (R^2),

$$R^2 = 1 - \frac{\sum_{l=1}^{t}(P_l - Q_l)^2}{\sum_{l=1}^{t}(P_l)^2} \tag{8}$$

where P is the desired value, Q is the network output value, and t is the number of samples.

4. Results and Discussion

For the RBF neural network, the Spread is usually a very important factor influencing the training process. Figure 4 shows the relationships of the mean square error (MSE) and the number of hidden layer neurons with different Spreads. Analyzing the results reported in Figure 4, it is found that for the same converged target, the neuron numbers in the hidden layer need to be increased obviously with the decrease of the Spread. When the Spread varies from 1 to 0.1, the corresponding neuron configuration of the RBF neural network are 5-38-1, 5-40-1, 5-56-1, 5-67-1, and 5-105-1, respectively. With the decrease of the Spread, the CPU time for computing the RBF neural network will also increase. At a Spread of 1, 0.5, 0.3, 0.2, and 0.1, the corresponding CPU times are 6.318, 6.396, 8.798, 10.827, and 15.772 s, respectively. In addition, Table 2 lists the values of four evaluation criteria for predicting the viscosity ratio of ethylene glycol/water based nanofluids by using the RBF neural network with different Spreads. It can be seen from Table 2 that although all R^2 are within the acceptable level of 0.99, the prediction performance of the RBF neural network is still affected by the value of the Spread, especially for the testing samples. Based on the comprehensive considerations of modeling complexity, prediction accuracy, and CPU time, the RBF neural network with the neuron configuration of 5-56-1 and Spread of 0.3 is used in this study. The related weights and biases of the 5-56-1 RBF neural network can be found in Table 3.

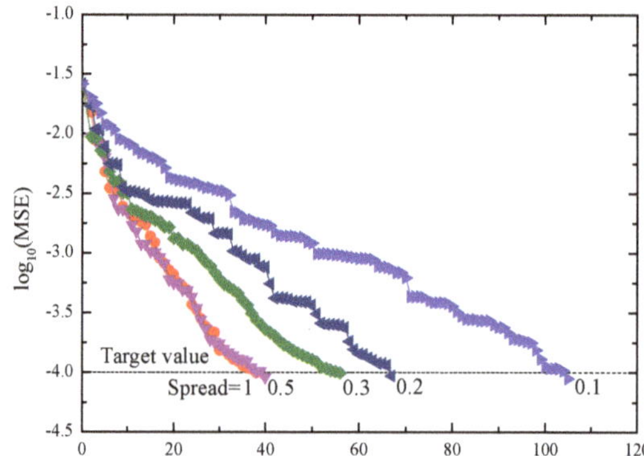

Figure 4. The relationships of mean square error (MSE) and the number of hidden layer neurons with different Spreads.

Table 2. Performance evaluation of RBF neural networks with different Spreads.

Object	Evaluation Criteria	Spread				
		1	0.5	0.3	0.2	0.1
Training samples	*RMSE*	0.04651	0.04460	0.04630	0.04502	0.04400
	MAPE (%)	2.3335	2.27321	2.09967	1.93784	2.00530
	SSE	0.42829	0.39383	0.42454	0.40125	0.38337
	R^2	0.99925	0.99932	0.99927	0.99930	0.99934
Testing samples	*RMSE*	0.09190	0.07035	0.04443	0.05124	0.10061
	MAPE (%)	4.65795	3.90042	2.43228	2.67720	4.27471
	SSE	0.15201	0.08908	0.03553	0.04727	0.18221
	R^2	0.99590	0.99760	0.99904	0.99872	0.99508
Total samples	*RMSE*	0.05183	0.04728	0.04615	0.04557	0.05117
	MAPE (%)	2.52721	2.40881	2.12738	1.99945	2.19442
	SSE	0.58030	0.48291	0.46007	0.44852	0.56558
	R^2	0.99906	0.99921	0.99925	0.99927	0.99908

Table 3. Weight and bias coefficients of the developed RBF neural network.

Neuron	Hidden Layer						Output Layer	
	Weights (w_{ij}) [a] and Biases						Weights (w_{ij}) [b] and Biases	
	T	d_p	ϕ_p	ρ_p	μ_f	Biases	μ_{nf}/μ_f	Biases
1	0.1616	0.6000	0.5952	1.0000	0.0817	2.7752	1.5072	0.0163
2	−0.4874	0.6000	0.7143	1.0000	0.6859	2.7752	2.5903	
3	0.6452	0.5660	0.3214	0.4200	0.0973	2.7752	0.0601	
4	0.8200	0.6000	0.7143	1.0000	0.0233	2.7752	3.4089	
5	0.4080	0.6000	0.0119	0.0174	0.1053	2.7752	−0.4782	
6	−0.1629	0.6000	0.7143	1.0000	0.2041	2.7752	24.6321	
7	0.4080	0.5660	0.9881	0.4200	0.1861	2.7752	−0.1055	
8	0.7818	0.2000	0.0119	0.4200	0.0211	2.7752	−0.0298	
9	−0.5716	0.6000	0.5952	1.0000	1.0000	2.7752	−2.4385	
10	0.4892	0.6000	0.1190	1.0000	0.0403	2.7752	0.0877	
11	0.3270	0.6000	0.7143	1.0000	0.0559	2.7752	2.0103	
12	−0.1629	0.6000	0.1190	1.0000	0.2041	2.7752	−16.7539	
13	0.8159	0.6000	0.0595	0.0174	0.0925	2.7752	0.2123	
14	0.3264	0.6000	0.0595	0.0174	0.1075	2.7752	1.5469	
15	0.4080	0.9120	0.4762	0.4200	0.1861	2.7752	0.1479	
16	0.9620	0.5660	0.5714	0.4200	0.0410	2.7752	0.2334	
17	0.4080	0.3620	0.1548	0.4200	0.1861	2.7752	−0.0212	
18	−0.1629	0.6000	0.5952	1.0000	0.2041	2.7752	−75.1746	
19	−0.4905	0.6000	0.1190	1.0000	0.6952	2.7752	−0.0522	
20	−0.4874	0.6000	0.5952	1.0000	0.6859	2.7752	−3.3888	
21	0.9819	0.5000	0.0476	0.6704	0.0389	2.7752	0.0377	
22	−0.1629	0.6000	0.4762	1.0000	0.2041	2.7752	115.2566	
23	0.6540	1.0000	0.7143	1.0000	0.0301	2.7752	1.0610	
24	0.6540	0.9000	0.7143	1.0000	0.0301	2.7752	−1.1272	
25	−0.3251	0.6000	0.5952	1.0000	0.3582	2.7752	0.2584	
25	−0.5685	0.6000	0.7143	1.0000	0.9855	2.7752	2.1391	
26	−0.5716	0.6000	0.4762	1.0000	1.0000	2.7752	1.2104	
27	0.4080	0.3620	1.0000	0.4200	0.1861	2.7752	0.4979	
28	0.4080	0.9120	1.0000	0.4200	0.1861	2.7752	0.2536	
29	0.6562	0.5000	0.2143	0.6704	0.0945	2.7752	−0.0974	
30	0.3270	0.6000	0.5952	1.0000	0.0559	2.7752	−3.6703	
31	0.4080	0.9120	0.0714	0.4200	0.1861	2.7752	0.1275	
32	0.4080	0.3620	0.4762	0.4200	0.1861	2.7752	0.2454	
33	0.4553	0.2000	0.0119	0.4200	0.0172	2.7752	0.1747	
34	1.0000	0.2000	0.0119	0.4200	0.0161	2.7752	0.1852	
35	0.1616	0.6000	0.4762	1.0000	0.0817	2.7752	−0.5112	
36	0.8169	0.6000	0.1190	1.0000	0.0234	2.7752	0.4325	
37	−0.4874	0.6000	0.4762	1.0000	0.6859	2.7752	1.6740	
38	−0.1629	0.6000	0.3571	1.0000	0.2041	2.7752	−107.8889	
39	0.3270	0.6000	0.4762	1.0000	0.0559	2.7752	1.4652	
40	0.1632	0.6000	0.0595	0.0174	0.1139	2.7752	0.9067	
41	0.8169	0.6000	0.5952	1.0000	0.0234	2.7752	−6.2589	
42	0.8169	0.6000	0.4762	1.0000	0.0234	2.7752	5.5200	
43	0.5739	0.5000	0.0119	0.6704	0.1182	2.7752	0.0349	
44	0.4080	0.3620	0.8095	0.4200	0.1861	2.7752	−0.1272	
45	0.4892	0.6000	0.5952	1.0000	0.0403	2.7752	0.6498	
46	−0.1629	0.6000	0.2381	1.0000	0.2041	2.7752	61.3817	
47	0.3283	0.5000	0.0119	0.6704	0.2312	2.7752	0.0577	
48	0.4080	0.5660	0.3095	0.4200	0.1861	2.7752	−0.1083	
49	0.2448	0.6000	0.0595	0.0174	0.1096	2.7752	−1.7397	
50	−0.4855	0.9000	0.7143	1.0000	0.6804	2.7752	−0.7083	
51	−0.4886	1.0000	0.7143	1.0000	0.6896	2.7752	0.6218	
52	−0.0007	0.6000	0.7143	1.0000	0.1252	2.7752	−0.7176	
53	−0.5692	0.9000	0.7143	1.0000	0.9885	2.7752	−0.0385	
54	0.6514	0.6000	0.2381	1.0000	0.0302	2.7752	−0.1993	
55	0.8169	0.6000	0.3571	1.0000	0.0234	2.7752	−2.0841	
56	0.1616	0.6000	0.5952	1.0000	0.0817	2.7752	1.5072	

[a] Weight connection from the input layer to the hidden layer; [b] Weight connection from the hidden layer to the output layer.

Figure 5 compares the predicted viscosity ratio of the RBF neural network and the experimental data involving the training and testing samples. It can be seen that all the prediction errors of the RBF neural network are within the ±10% error bands. As shown in Table 2, the values of the four evaluation criteria are $RMSE$ = 0.04630, $MAPE$ = 2.09967%, SSE = 0.42454, and R^2 = 0.99927 for the training samples, and $RMSE$ = 0.04443, $MAPE$ = 2.43228%, SSE = 0.03553, and R^2 = 0.99904 for the testing samples, which preliminarily indicates that the RBF neural network has a good ability to predict the viscosity ratio of ethylene glycol/water based nanofluids.

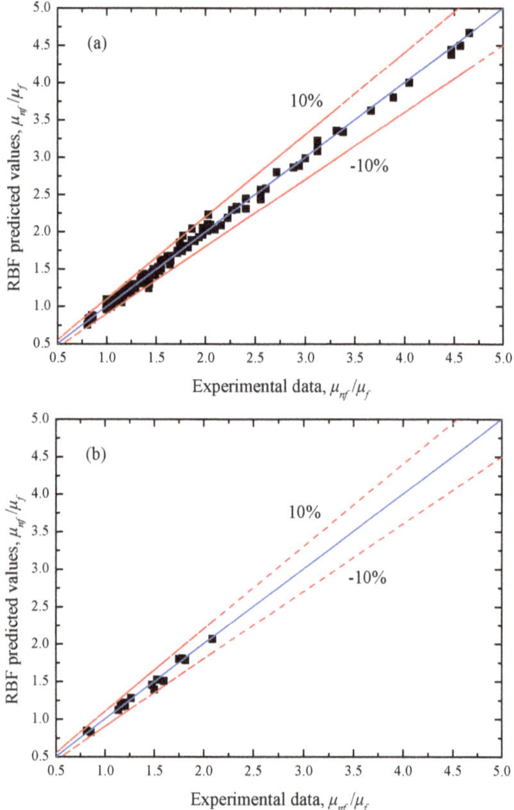

Figure 5. Scatter plots of (**a**) training and (**b**) testing μ_{nf}/μ_f for the RBF predicted results and experimental data.

To further evaluate the prediction performance of the RBF neural network for nanofluid viscosity, the following typical viscosity models which consider the effects of nanoparticle Brownian motion and aggregation are selected for analysis.

Batchelor model [32]:

$$\mu_{nf} = (1 + 2.5\phi_p + 6.25\phi_p^2)\mu_f \tag{9}$$

Chen model [33]:

$$\mu_{nf} = [1 - \frac{\phi_p}{0.605}(\frac{r_a}{r_p})^{1.2}]^{-1.5125}\mu_f \tag{10}$$

where r_p and r_a are the radius of nanoparticle and nanoparticle aggregation, respectively.

Figure 6 and Table 4 respectively compare the prediction performances of the different models for the total viscosity data. It is easily seen that the RBF neural network has a better prediction accuracy than the above two typical models. The main reason is that the Batchelor model and Chen model cannot fully quantitatively describe the relationship between the nanofluid viscosity ratio and the various factors including the nanoparticle properties, temperature, and base fluid.

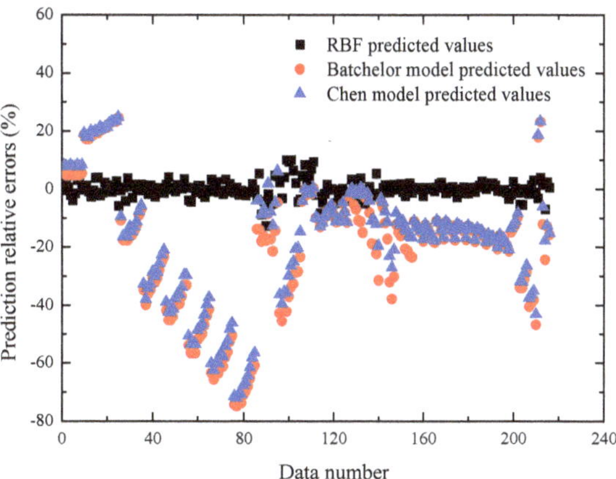

Figure 6. Prediction relative errors of different models for the total viscosity data.

Table 4. Performance evaluation of different modes for the total viscosity data.

Evaluation Criteria	RBF Neural Network	Batchelor Model	Chen Model
RMSE	0.04615	0.82200	0.77129
MAPE (%)	2.12738	24.2349	21.79539
SSE	0.46007	145.94871	128.49640
R^2	0.99925	0.76244	0.79085

Moreover, Tables 5–7 respectively present the comparisons between the predicted viscosity ratio of the RBF neural network and the corresponding experimental data of Chen et al. [12], Jamshidi et al. [13], and Namburu et al. [16]. It can be seen that there is good agreement between the RBF predicted and the experimental viscosity ratio of the different ethylene glycol/water based nanofluids. At the temperature range of 20–40 °C, the maximum and minimum prediction errors of the RBF neural network are respectively 5.788% and 0.434% for the experimental data of Chen et al. [12]. For the viscosity ratio of the SiO_2-ethylene glycol/water (50:50 by volume ratio) nanofluid provided by Jamshidi et al. [13], the RBF neural network can accurately predict the viscosity ratio with an average error of 1.772% at the nanoparticle volume fraction of 0.1%. Moreover, the comparisons shown in Table 7 further illustrate that the developed RBF neural network has high accuracy (average error: 2.097%) for predicting the viscosity ratio of CuO-ethylene glycol/water (60:40 by weight ratio) nanofluids.

Table 5. Comparisons of the RBF predicted viscosity ratio of TiO_2-ethylene glycol nanofluids with the experimental data of Chen et al. [12].

T (°C)	ϕ_p (%)	Experiment (P)	RBF Prediction (Q)	$\frac{\|P-Q\|}{P}\times 100\%$
60.17	1.8	0.994	0.990	0.434
55.26	1.8	1.000	1.023	2.330
50.03	1.8	1.000	1.005	0.488
45.27	1.8	0.994	0.974	2.090
40.21	1.8	1.000	0.961	3.875
35.17	1.8	1.000	0.985	1.529
30.16	1.8	0.994	1.025	3.055
25.19	1.8	1.000	1.041	4.068
20.12	1.8	0.994	0.995	0.082
60.17	0.4	0.852	0.850	0.270
55.10	0.4	0.862	0.872	1.217
50.03	0.4	0.862	0.855	0.829
45.12	0.4	0.857	0.830	3.173
40.07	0.4	0.847	0.825	2.603
35.17	0.4	0.852	0.848	0.424
30.16	0.4	0.847	0.880	3.868
25.19	0.4	0.847	0.881	4.038
20.12	0.4	0.837	0.819	2.193
60.17	0.1	0.833	0.822	1.268
55.10	0.1	0.828	0.846	2.220
50.19	0.1	0.828	0.833	0.609
45.12	0.1	0.819	0.810	1.011
40.07	0.1	0.814	0.806	1.020
35.17	0.1	0.814	0.824	1.284
30.16	0.1	0.814	0.847	4.039
25.19	0.1	0.814	0.836	2.675
20.12	0.1	0.805	0.758	5.788

Table 6. Comparisons of the RBF predicted viscosity ratio of SiO_2-ethylene glycol/water (50:50 by volume ratio) nanofluids with the experimental data of Jamshidi et al. [13].

T (°C)	ϕ_p (%)	Experiment (P)	RBF Prediction (Q)	$\frac{\|P-Q\|}{P}\times 100\%$
28.65	0.1	1.082	1.099	1.549
38.18	0.1	1.076	1.073	0.258
47.91	0.1	1.064	1.063	0.067
55.81	0.1	1.100	1.098	0.202
61.28	0.1	1.132	1.078	4.746
28.45	0.1	1.128	1.098	2.703
36.55	0.1	1.119	1.082	3.323
45.07	0.1	1.054	1.056	0.196
50.95	0.1	1.091	1.078	1.243
58.85	0.1	1.134	1.095	3.432

Table 7. Comparisons of the RBF predicted viscosity ratio of CuO-ethylene glycol/water (60:40 by weight ratio) nanofluids with the experimental data of Namburu et al. [16].

T (°C)	ϕ_p (%)	Experiment (P)	RBF Prediction (Q)	$\frac{\|P-Q\|}{P}\times 100\%$
9.970	1.000	1.204	1.181	1.940
20.305	1.000	1.187	1.217	2.547
29.860	1.000	1.170	1.174	0.308
40.196	1.000	1.136	1.123	1.195
0.006	2.000	1.596	1.511	5.307
10.146	2.000	1.596	1.508	5.504
20.288	2.000	1.528	1.528	0.019
30.040	2.000	1.477	1.459	1.216
−0.199	3.000	1.817	1.789	1.566
10.137	3.000	1.783	1.809	1.451
20.082	3.000	1.749	1.803	3.085
29.620	4.000	2.089	2.068	1.028

Figure 7 compares the experimental viscosity ratio of Rudyak et al. [15] with the predicted values of the RBF neural network for the SiO_2-ethylene glycol nanofluids at T = 25 °C as a function of the nanoparticle volume fraction and diameter. It can be found from Figure 7a that the RBF predicted viscosity ratio of nanofluids are obviously enhanced with the increase of the SiO_2 nanoparticle volume fraction and the decrease of the nanoparticle size, which are consistent with the experimental results. All the prediction relative errors are within ±8%, as shown in Figure 7b. On this basis, Figure 8 illustrates the comparisons between the RBF predicted values and the corresponding experimental data of Li et al. [19]. The results indicate that the RBF neural network developed in this study can be applied successfully for predicting the effects of the nanoparticle volume fraction and temperature on the viscosity ratio of SiC-ethylene glycol/water (40:60 by weight ratio) nanofluids with a satisfactory accuracy. In addition, a similar analysis is performed for the CuO-ethylene glycol/water (60:40 by weight ratio) nanofluids as a function of temperature, which is presented in Figure 9. It is demonstrated that the viscosity ratio characteristics of the above nanofluids are effectively predicted by the RBF neural network in a wide range of nanoparticle volume fractions (from 1% to 6%) and temperatures (from −35 to 50 °C). The maximum prediction relative errors are only 4.2%. All the above analyses further demonstrate that the RBF neural network is one of the potential tools to quantitatively establish nonlinear relationships between inputs and outputs.

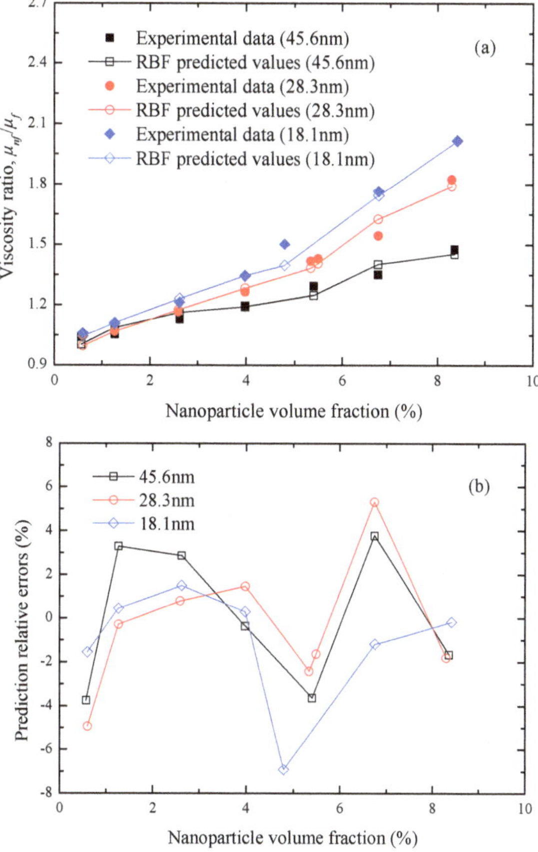

Figure 7. (**a**) Predicted comparisons and (**b**) relative errors of the RBF predicted μ_{nf}/μ_f and the experimental data [15] for SiO_2-ethylene glycol nanofluids at T = 25 °C.

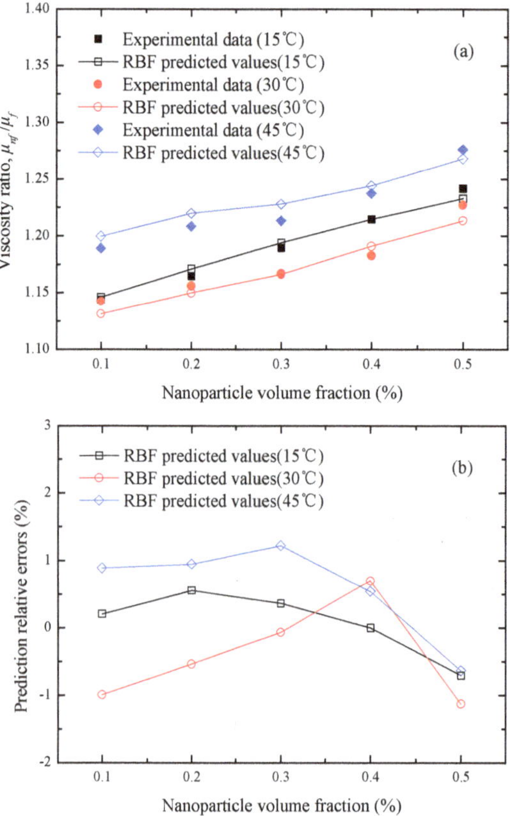

Figure 8. (**a**) Predicted comparisons and (**b**) relative errors of the RBF predicted μ_{nf}/μ_f and the experimental data [19] for SiC-ethylene glycol/water (40:60 by weight ratio) nanofluids at d_p = 30 nm.

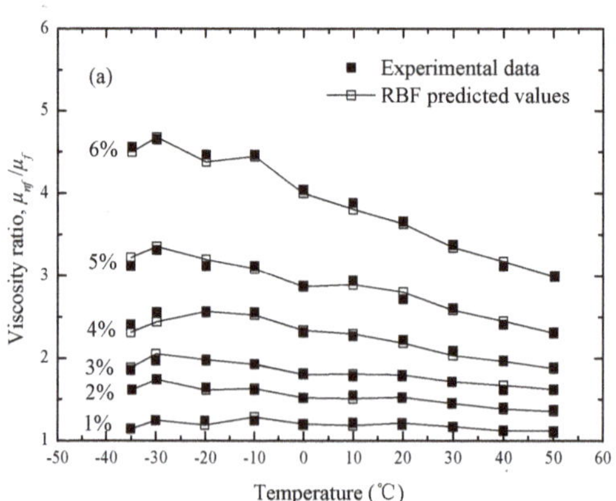

Figure 9. *Cont.*

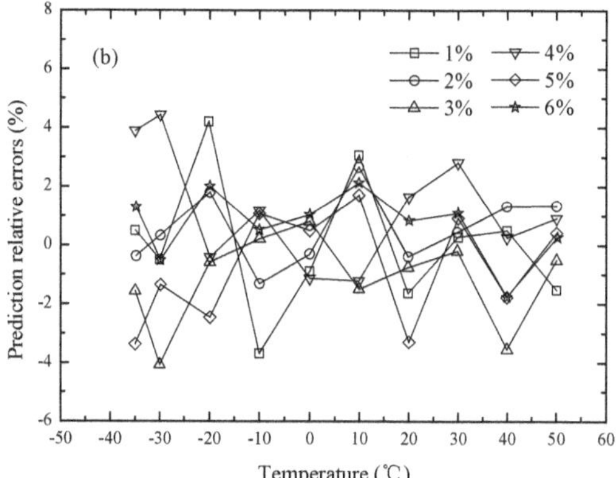

Figure 9. (**a**) Predicted comparisons and (**b**) relative errors of the RBF predicted μ_{nf}/μ_f and the experimental data [14] for CuO-ethylene glycol/water (60:40 by weight ratio) nanofluids at d_p = 30 nm.

Table 8 shows the prediction performance of the RBF neural network using different viscosity data sets. It is worth noting that the data sets are selected randomly. From Table 8, we found that the size of the data set can affect the modeling and prediction of the RBF neural network significantly. With the decrease of the data set size, the prediction accuracy will decrease. This may mean that to accurately predict the viscosity of ethylene glycol/water based nanofluids using the RBF neural network, a large enough data set is necessary.

Table 8. Performance evaluation of the RBF neural network with different viscosity data.

Evaluation Criteria	216 Data	200 Data	160 Data	120 Data
RMSE	0.04615	0.07017	0.09547	0.37982
MAPE (%)	2.12738	2.88920	3.64470	8.75159
SSE	0.46007	0.98468	1.45844	17.31116
R^2	0.99925	0.99832	0.99505	0.90323

5. Conclusions

To accurately predict the viscosity ratio between ethylene glycol/water nanofluids and a base fluid, a RBF neural network based model was developed and evaluated in the present study. Based on the comparative analysis, the following conclusions were obtained.

(1) Considering the complex effects of different factors including temperature, nanoparticle properties (such as volume fraction, density, diameter), and viscosity of the base fluid on the viscosity ratio and the effect of Spread on modeling performance of the RBF neural network, the final network structure was determined to be 5-56-1 neurons.

(2) By comparing the RBF predictive values and the experimental data published in various studies, it was demonstrated that the RBF neural network not only exhibited good modeling accuracy (*RMSE* = 0.04615, *MAPE* = 2.12738%, *SSE* = 0.46007, R^2 = 0.99925), but also could effectively predict the influences of temperature, nanoparticle volume fraction, and diameter on the viscosity ratio of different ethylene glycol/water based nanofluids.

(3) Compared to the typical viscosity models, namely the Batchelor model and Chen model, the RBF neural network has a good ability to predict the viscosity ratio of different ethylene glycol/water based nanofluids. However, the prediction performance can be affected by the size of the data set.

(4) The present investigation may play an active role for developing the modeling of nanofluid viscosity. However, how to extend the application of ANN to predict other thermo-physical properties of nanofluids is still worthy of study in the future.

Acknowledgments: The authors acknowledge the financial support by the Fundamental Research Funds for the Central Universities (No. HEUCF160307).

Author Contributions: Ningbo Zhao was responsible for the main parts of this manuscript, which includes the collection of viscosity data and the results analysis. Zhiming Li provided the program codes of RBF neural network and the conventional viscosity models, and wrote the basic theory of the RBF neural network.

Conflicts of Interest: We declare that we have no conflict of interest.

References

1. Peyghambarzadeh, S.M.; Hashemabadi, S.H.; Hoseini, S.M.; Jamnani, M.S. Experimental study of heat transfer enhancement using water/ethylene glycol based nanofluids as a new coolant for car radiators. *Int. Commun. Heat Mass* **2011**, *38*, 1283–1290. [CrossRef]
2. Garoosi, F.; Jahanshaloo, L.; Rashidi, M.M.; Badakhsh, A.; Ali, M.E. Numerical simulation of natural convection of the nanofluid in heat exchangers using a Buongiorno model. *Appl. Math. Comput.* **2015**, *254*, 183–203. [CrossRef]
3. Zhao, N.B.; Li, S.Y.; Yang, J.L. A review on nanofluids: Data-driven modeling of thermalphysical properties and the application in automotive radiator. *Renew. Sustain. Energy Rev.* **2016**, *66*, 596–616. [CrossRef]
4. Hussein, A.M.; Kadirgama, K.; Noor, M.M. Nanoparticles suspended in ethylene glycol thermal properties and applications: An overview. *Renew. Sust. Energy Rev.* **2016**, *69*, 1324–1330. [CrossRef]
5. Vajjha, R.S.; Das, D.K. Experimental determination of thermal conductivity of three nanofluids and development of new correlations. *Int. J. Heat Mass Transf.* **2009**, *52*, 4675–4682. [CrossRef]
6. Sundar, L.S.; Singh, M.K.; Sousa, A.C.M. Thermal conductivity of ethylene glycol and water mixture based Fe_3O_4 nanofluid. *Int. Commun. Heat Mass* **2013**, *49*, 17–24. [CrossRef]
7. Bég, O.A.; Rashidi, M.M.; Akbari, M.; Hosseini, A. Comparative numerical study of single-phase and two-phase models for bio-nanofluid transport phenomena. *J. Mech. Med. Biol.* **2014**, *14*, 1450011. [CrossRef]
8. Garoosi, F.; Rohani, B.; Rashidi, M.M. Two-phase mixture modeling of mixed convection of nanofluids in a square cavity with internal and external heating. *Powder Technol.* **2015**, *275*, 304–321. [CrossRef]
9. Sheikholeslami, M.; Rashidi, M.M.; Hayat, T.; Ganji, D.D. Free convection of magnetic nanofluid considering MFD viscosity effect. *J. Mol. Liq.* **2016**, *218*, 393–399. [CrossRef]
10. Azmi, W.H.; Hamid, K.A.; Mamat, R.; Sharma, K.V.; Mohamad, M.S. Effects of working temperature on thermo-physical properties and forced convection heat transfer of TiO_2 nanofluids in water–ethylene glycol mixture. *Appl. Therm. Eng.* **2016**, *106*, 1190–1199. [CrossRef]
11. Sundar, L.S.; Ramana, E.V.; Singh, M.K.; de Sousa, A.C.M. Viscosity of low volume concentrations of magnetic Fe_3O_4 nanoparticles dispersed in ethylene glycol and water mixture. *Chem. Phys. Lett.* **2012**, *554*, 236–242. [CrossRef]
12. Chen, H.; Ding, Y.; Tan, C. Rheological behaviour of nanofluids. *New J. Phys.* **2007**, *9*, 367. [CrossRef]
13. Jamshidi, N.; Farhadi, M.; Ganji, D.; Sedighi, K. Experimental investigation on viscosity of nanofluids. *Int. J. Eng.* **2012**, *25*, 201–209. [CrossRef]
14. Kulkarni, D.P.; Das, D.K.; Vajjha, R.S. Application of nanofluids in heating buildings and reducing pollution. *Appl. Energy* **2009**, *86*, 2566–2573. [CrossRef]
15. Rudyak, V.Y.; Dimov, S.V.; Kuznetsov, V.V. On the dependence of the viscosity coefficient of nanofluids on particle size and temperature. *Tech. Phys. Lett.* **2013**, *39*, 779–782. [CrossRef]
16. Namburu, P.K.; Kulkarni, D.P.; Misra, D.; Das, D.K. Viscosity of copper oxide nanoparticles dispersed in ethylene glycol and water mixture. *Exp. Therm. Fluid Sci.* **2007**, *32*, 397–402. [CrossRef]
17. Lim, S.K.; Azmi, W.H.; Yusoff, A.R. Investigation of thermal conductivity and viscosity of Al_2O_3/water–ethylene glycol mixture nanocoolant for cooling channel of hot-press forming die application. *Int. Commun. Heat Mass* **2016**, *78*, 182–189. [CrossRef]

18. Chiam, H.W.; Azmi, W.H.; Usri, N.A.; Mamat, R.; Adam, N.M. Thermal conductivity and viscosity of Al_2O_3 nanofluids for different based ratio of water and ethylene glycol mixture. *Exp. Therm. Fluid Sci.* **2017**, *81*, 420–429. [CrossRef]
19. Li, X.; Zou, C.; Qi, A. Experimental study on the thermo-physical properties of car engine coolant (water/ethylene glycol mixture type) based SiC nanofluids. *Int. Commun. Heat Mass* **2016**, *77*, 159–164. [CrossRef]
20. Murshed, S.S.; Estellé, P. A state of the art review on viscosity of nanofluids. *Renew. Sust. Energy Rev.* **2017**, *76*, 1134–1152. [CrossRef]
21. Heidari, E.; Sobati, M.A.; Movahedirad, S. Accurate prediction of nanofluid viscosity using a multilayer perceptron artificial neural network (MLP-ANN). *Chemom. Intell. Lab.* **2016**, *155*, 73–85. [CrossRef]
22. Yousefi, F.; Karimi, H.; Papari, M.M. Modeling viscosity of nanofluids using diffusional neural networks. *J. Mol. Liq.* **2015**, *175*, 85–90. [CrossRef]
23. Mehrabi, M.; Sharifpur, M.; Meyer, J.P. Viscosity of nanofluids based on an artificial intelligence model. *Int. Commun. Heat Mass* **2013**, *43*, 16–21. [CrossRef]
24. Zhao, N.B.; Li, S.Y.; Wang, Z.T.; Cao, Y.P. Prediction of viscosity of nanofluids using artificial neural networks. In Proceedings of the ASME 2014 International Mechanical Engineering Congress & Exposition, Montreal, QC, Canada, 14–20 November 2014.
25. Zhao, N.B.; Wen, X.Y.; Yang, J.L.; Li, S.Y.; Wang, Z.T. Modeling and prediction of viscosity of water-based nanofluids by radial basis function neural networks. *Powder Technol.* **2015**, *281*, 173–183. [CrossRef]
26. Yang, S.; Cao, Y.; Peng, Z.; Wen, G.; Guo, K. Distributed formation control of nonholonomic autonomous vehicle via RBF neural network. *Mech. Syst. Signal Process.* **2017**, *87*, 81–95. [CrossRef]
27. Broomhead, D.S.; Lowe, D. Radial basis functions, multi-variable functional interpolation and adaptive networks. *Complex Syst.* **1988**, *2*, 321–355.
28. Li, M.M.; Verma, B. Nonlinear curve fitting to stopping power data using RBF neural networks. *Expert Syst. Appl.* **2016**, *45*, 161–171. [CrossRef]
29. Turnbull, D.; Elkan, C. Fast recognition of musical genres using RBF networks. *IEEE Trans. Knowl. Data Eng.* **2005**, *17*, 580–584. [CrossRef]
30. Iliyas, S.A.; Elshafei, M.; Habib, M.A.; Adeniran, A.A. RBF neural network inferential sensor for process emission monitoring. *Control Eng. Pract.* **2013**, *21*, 962–970. [CrossRef]
31. Chen, S.; Cowan, C.F.N.; Grant, P.M. Orthogonal least squares learning algorithm for radial basis function networks. *IEEE Trans. Neural Netw.* **1991**, *2*, 302–309. [CrossRef] [PubMed]
32. Batchelor, G.K. The effect of Brownian motion on the bulk stress in a suspension of spherical particles. *J. Fluid Mech.* **1977**, *83*, 97–117. [CrossRef]
33. Chen, H.; Ding, Y.; He, Y.; Tan, C. Rheological behaviour of ethylene glycol based titania nanofluids. *Chem. Phys. Lett.* **2007**, *444*, 333–337. [CrossRef]

Article

Convective Heat Transfer and Particle Motion in an Obstructed Duct with Two Side by Side Obstacles by Means of DPM Model

Saman Rashidi [1,†], Javad Aolfazli Esfahani [2,†] and Rahmat Ellahi [3,*,†]

1. Department of Mechanical Engineering, Semnan Branch, Islamic Azad University, Semnan, Iran; samanrashidi3983@gmail.com
2. Department of Mechanical Engineering, Ferdowsi University of Mashhad, Mashhad 91775-1111, Iran; abolfazl@um.ac.ir
3. Department of Mathematics & Statistics, FBAS, IIUI, Islamabad 44000, Pakistan

* Correspondence: rahmatellahi@yahoo.com; Tel.: +92-51-9019-510

† All authors contributed equally to this work.

Academic Editors: Yulong Ding and Yuyuan Zhao

Received: 11 January 2017; Accepted: 19 April 2017; Published: 24 April 2017

Abstract: In this research, a two-way coupling of discrete phase model is developed in order to track the discrete nature of aluminum oxide particles in an obstructed duct with two side-by-side obstacles. Finite volume method and trajectory analysis are simultaneously utilized to solve the equations for liquid and solid phases, respectively. The interactions between two phases are fully taken into account in the simulation by considering the Brownian, drag, gravity, and thermophoresis forces. The effects of space ratios between two obstacles and particle diameters on different parameters containing concentration and deposition of particles and Nusselt number are studied for the constant values of Reynolds number ($Re = 100$) and volume fractions of nanoparticles ($\Phi = 0.01$). The obtained results indicate that the particles with smaller diameter ($d_p = 30$ nm) are not affected by the flow streamline and they diffuse through the streamlines. Moreover, the particle deposition enhances as the value of space ratio increases. A comparison between the experimental and numerical results is also provided with the existing literature as a limiting case of the reported problem and found in good agreement.

Keywords: concentration; deposition; two-way coupling; side-by-side obstacles; discrete phase model (DPM)

1. Introduction

Investigations of the momentum and heat transfer specifications in an obstructed duct with multiple obstacles has many thermal applications including compact heat exchangers, flow around arrays of nuclear fuel rods, cooling of electronic devices, oil or gas flows in reservoirs, chimney stacks, power generators, etc. [1]. The heat transfer improvement in all mentioned applications is an essential need. Nanofluids are recognized as high heat transfer performance fluids, which can be used in many cooling systems including cooling of electronic components, oil coolers, inter coolers, and coolant in microchannel heat sink [2,3]. The particulate fouling as a destructive phenomenon in nanofluids should be taken into consideration as it can affect the favorable improved thermal properties of these fluids. Accurate understanding of the particle motion is essential to supress the destructive effects of this phenomenon.

In past years, researchers used different models for simulating nanofluid flow for different problems. These models are single-phase approach, Eulerian model, volume of fluid model, mixture model, and Eulerian–Lagrangian approach (Discrete phase model). Vanaki et al. [4] reviewed these models. They concluded that the Eulerian-Lagrangian approach is more precise and reliable for simulating the nanofluid because it takes into account the interactions between two phases by considering the Brownian, drag, gravity, and thermophoresis forces between them. Moreover, this model has ability to predict particle distribution and calculate concentration of nanoparticles in domain. He et al. [5] performed a comparison between single-phase and Lagrangian trajectory approaches for simulating the nanofluid flow in a tube. They found that the Lagrangian trajectory approach predicts a higher heat transfer coefficient in comparison with the single-phase one as the Lagrangian trajectory approach considers the dynamic of particles and the interactions between liquid and particulate phases. Mirzaei et al. [6] used the Eulerian–Lagrangian approach to investigate the nanofluid flow in a microchannel. They reported the same findings about the greater prediction of heat transfer coefficient by this model in comparison to the single-phase one. In another study, Bahremand et al. [7] studied the nanofluid turbulent flow in helically-coiled tubes. They used the Eulerian-Lagrangian approach. Besides the numerical study, they performed an experimental work on this problem to benchmark the accuracy of the numerical model. They observed that the Eulerian–Lagrangian model presents the results with higher accuracy in comparison to the single-phase one. Some researchers studied the flow and heat transfer in an obstructed channel. Turki et al. [8] studied the convective heat transfer across an obstructed duct with a built-in heated square obstacle. Their results indicated that the Strouhal number increases with an enhance in the Richardson number. Mohammadi Pirouz et al. [9] simulated the heat exchange in a duct with wall-mounted square obstacles. They reported that the flow accelerates near faces with a decrease in the distance between obstacles. This leads to enhance in the heat transfer rate from obstacles. Heidary and Kermani [10] investigated the influences of nanofluid on heat transfer improvement in an obstructed duct. They applied the single-phase approach for modelling the nanofluid for this problem. They observed about 60% enhancement for heat transfer in the duct by using the nanoparticles and the blocks. Readers are referred to the most significant studies on nanoparticles in [11–20] and several references therein. Recently, Shahmohamadi and Rashidi [21] analytically studied how nanofluids flow through a rotating channel with a lower stretching porous wall under the influence of a magnetic field. They reported that the nanoparticle additives have considerable influence on the flow. In other research, Shahmohamadi et al. [22] investigated tribological performance of carbon nanoparticles dispersed in polyalphaolefin PAO6 oil. They concluded that the presence of nanoparticles causes a higher lubricant viscosity.

Previous researchers concluded that the Eulerian–Lagrangian model (Discrete phase model) is a superior model to simulate the nanofluid, as it can predict particle distribution and calculate concentration and deposition of nanoparticles in the domain. Enhancement of heat transfer is a very important problem for an obstructed duct with multiple obstacles as it has many thermal applications such as compact heat exchangers. Nanofluids can be introduced as an option to achieve this target. The particulate fouling phenomenon in nanofluids should be taken into consideration as it can affect the favorable improved thermal properties of these fluids. This paper simulates the convective heat transfer and particle motion and deposition in an obstructed duct with two side-by-side obstacles by a two-way coupling of DPM model. The present research represents the first study about the application of two-way coupling of DPM model to simulate nanofluid in this geometry.

2. Formulation of the Problem

2.1. Physical Characteristic

A view of the computational domain is disclosed in Figure 1. A two-dimensional obstructed duct with two side-by-side obstacles is modelled. Computational domain is subjected to a free stream with parabolic speed and uniform temperature (T_h = 310 K). It is assumed that the stream is incompressible, time dependent, and laminar. The obstacles and duct walls are kept at constant temperature (T_c = 300 K). Two side-by-side obstacles with sides D are mounted in the duct. S is the gap between the centers of the obstacles. The upstream and downstream lengths of the duct are fixed as 30D and 60D, respectively. The duct height is considered to be 16D. The gravity is in the stream-wise direction.

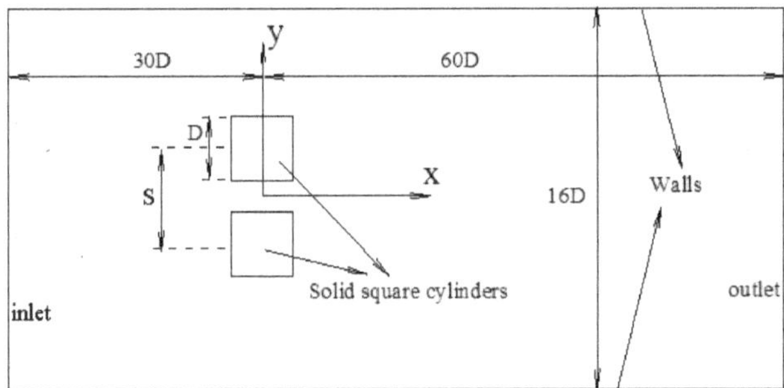

Figure 1. Computational domain and coordinate system.

The interactions between the liquid and aluminum oxide particles are taken into account for the simulations. Moreover, the collisions between particles are ignored. Note that for particles in the range of nanometer with low values of volume fraction, the chances that any two random particles would be close enough to interact with each other are extremely low [23]. Accordingly, the interactions between the particles in the range of nanometer can be ignored. However, for particles with higher diameter (e.g., 0.5 μm), it is better to consider the effects of particle-particle interactions. It should be stated that in DPM, it is not possible to track all physical particles. Instead, representative particles or parcels should be tracked. Any parcel is representative of specified number of actual particles with the same physical property, which is characterized by the particle flow rate along each calculated particle trajectory. 5000 parcels [24] are assumed at $\Phi = 0.01$ for this research. Finally, it should be stated that all simulations are performed for the fixed values of Reynolds number ($Re = 100$) and volume fractions of nanoparticles ($\Phi = 0.01$) at variable particle diameter (30 nm–0.5 μm) and space ratio ($S/D = 1.5$–4.5).

2.2. Governing Equations

For two-way coupling of Eulerian-Lagrangian approach, the liquid phase is treated as a continuum by using the Eulerian approach, while the dispersed particles can be tracked by applying a Lagrangian approach. The heat, mass, and momentum can be exchanged between two phases [25,26]. The equations for two phases are presented separately as follows:

2.2.1. Liquid Phase

- Mass conservation equation:

$$\frac{\partial u}{\partial x} + \frac{\partial v}{\partial y} = 0 \tag{1}$$

where x and y are Cartesian coordinate components. Moreover, u and v are velocity components in x and y directions, respectively.

- Momentum equation:

$$\rho\left(\frac{\partial u}{\partial t} + u\frac{\partial u}{\partial x} + v\frac{\partial u}{\partial y}\right) = -\frac{\partial p}{\partial x} + \mu\left(\frac{\partial^2 u}{\partial x^2} + \frac{\partial^2 u}{\partial y^2}\right) + S_{vx} \tag{2}$$

$$\rho\left(\frac{\partial v}{\partial t} + u\frac{\partial v}{\partial x} + v\frac{\partial v}{\partial y}\right) = -\frac{\partial p}{\partial y} + \mu\left(\frac{\partial^2 v}{\partial x^2} + \frac{\partial^2 v}{\partial y^2}\right) + S_{vy} \tag{3}$$

where ρ, μ, p, and t are density of water, viscosity of water, pressure, and time, respectively.

- Energy equation:

$$\rho C_p\left(\frac{\partial T}{\partial t}+u\frac{\partial T}{\partial x}+v\frac{\partial T}{\partial y}\right)=k\left(\frac{\partial^2 T}{\partial x^2}+\frac{\partial^2 T}{\partial y^2}\right)+S_h \tag{4}$$

where C_p, k, and T are heat capacity of water, heat conductivity of water, and temperature, respectively. Moreover, the terms of S_v and S_h show the momentum and heat exchanges between two phases, respectively, and can be defined as [27,28]

$$S_v=\sum_{np}-\frac{m_p}{\delta V}\frac{d\,V_p}{dt} \tag{5}$$

$$S_h=\sum_{np}\frac{m_p}{\delta V}C_p\frac{dT_p}{dt} \tag{6}$$

where parameters with subscript of "p" are related to the particle phase. Accordingly, m_p, n_p, and δV are the particle mass, the number of particles in a cell volume, and the cell volume, respectively.

2.2.2. Particle Phase

For each particle suspended in the liquid, a differential form of force balance equation is utilized to obtain the trajectory of solid phase. The interaction forces between fluid and solid phases contain the drag, Brownian, gravity, and thermophoresis forces. The dynamic equations of solid phase are

$$\frac{dX_p}{dt}=V_p \tag{7}$$

$$\begin{aligned}\frac{dV_p}{dt}=&\underbrace{\frac{18\mu_f}{d_p^2\rho_p C_c}(V_f-V_p)}_{\text{Drag force}}+\underbrace{\frac{g(\rho_p-\rho_f)}{\rho_p}}_{\text{Gravitational force}}+\underbrace{\varsigma\sqrt{\frac{\pi S_0}{\Delta t}}}_{\text{Brownian force}}\\&-\underbrace{\frac{36\mu^2 C_s(k_f/k_p+C_t+Kn)}{\rho_f\rho_p d_p^2(1+3C_m Kn)(1+2k_f/k_p+2C_t Kn)}\frac{\nabla T}{T}}_{\text{Thermophoretic force}}\end{aligned} \tag{8}$$

where subscripts of "f" and "p" demonstrate the liquid and solid phases, respectively. X and V indicate the location and velocity of the particles, respectively. Moreover, g, k, t, ρ, and μ are gravitational acceleration, heat conductivity, time, density, and viscosity, respectively.

C_c in drag force term denotes the Cunningham correction and is determined by

$$C_c=1+\frac{2\lambda}{d_p}\left[1.257+0.4e^{-(1.1d/2\lambda)}\right] \tag{9}$$

where d_p and λ are the particle diameter and fluid mean free path, respectively. Moreover, Kn in thermophoresis force terms is the Knudsen number. The constant values of C_m, C_t, and C_s, are respectively allocated as 1.14, 2.18, and 1.17 [29]. ς in Brownian force is the zero-mean, unit-variance independent Gaussian random numbers. Finally, S_0 is the spectral intensity of Brownian force and is calculated as follows [30,31]:

$$S_0=\frac{216\nu K_B T_f}{\pi^2\rho_f d_p^5\left(\frac{\rho_p}{\rho_f}\right)^2 C_c} \tag{10}$$

where K_B is the Boltzmann constant (=1.38×10^{-23} J K^{-1}).

The energy equation for the solid phase is presented as

$$m_p C_p\left(\frac{dT_p}{dt}\right)=hA_p(T_f-T_p) \tag{11}$$

where A_p is the particle surface area. Moreover, h is the heat transfer coefficient and is defined by [32]:

$$h = \frac{k_f}{d_p}\left(2 + 0.6Re_p^{0.5}Pr_f^{0.3}\right) \tag{12}$$

where Pr_f and Re_p are the Prandtl number of the liquid phase and particle Reynolds number, respectively.

3. Boundary Conditions

3.1. Liquid Phase

At the entrance of the duct, a parabolic velocity and a uniform temperature are used for velocity and temperature fields, respectively. Accordingly, the boundary condition at this section is defined by

$$u = U_0(1 - (2y/D)^2),\ v = 0, T = T_h \tag{13}$$

where U_0 is the velocity at the center of the duct.

Along the surfaces of the duct and obstacles, no-slip condition and constant temperature are considered for velocity and temperature fields, respectively. The boundary conditions for these regions are expressed by

$$u = 0,\ v = 0,\ T = T_c \tag{14}$$

At exit of the duct, zero gradient boundary conditions are used for both velocity and temperature fields. The boundary conditions for this section are expressed by

$$\frac{\partial u}{\partial x} = 0,\ \frac{\partial v}{\partial x} = 0,\ \frac{\partial T}{\partial x} = 0 \tag{15}$$

Eventually, it is supposed that there is no flow across the duct and the temperature is constant at initial time. This condition can be introduced by

$$u = 0,\ v = 0,\ T = T_h \tag{16}$$

3.2. Solid Phase

In this research, the escape boundary is considered at the entrance and exit sections of the duct. Based on this boundary, the trajectory computations are stopped when a particle exits from the domain. The temperature of particles in the entrance section is fixed at 310 K and the particle temperature drops to 300 K by colliding the particles with a surface. Eventually, the reflect boundary with a restitution coefficient of 1 is considered for all surfaces containing the surfaces of the duct and obstacle.

3.3. Physical Parameter

Physical parameters involved in this study are introduced in this section.

The local surface Nusselt number based on the channel width is calculated by

$$Nu = \frac{hH}{k_f} = \left.\frac{\partial T^*}{\partial n^*}\right|_{\text{on obstacle}} \tag{17}$$

where n and H are the normal direction to the channel walls and the width of the channel, respectively. Note that k_f in the above equation is the thermal conductivity of the fluid. Also, superscript "*" denotes the non-dimensional variables. T^* and n^* are defined by

$$T^* = \frac{T - T_c}{T_m - T_c},\ n^* = \frac{n}{H} \tag{18}$$

where T_c is the temperature on the walls of the duct. Moreover, T_m is mean temperature, defined by

$$T_m = \frac{1}{Hu_m} \int_{-\frac{H}{2}}^{+\frac{H}{2}} uTdy \tag{19}$$

where u_m is the mean velocity, calculated by

$$u_m = \frac{1}{H} \int_{-\frac{H}{2}}^{+\frac{H}{2}} udy \tag{20}$$

The heat flux can be evaluated by

$$q''_{\text{on the wall}} = h(T_m - T_c) = k_f \frac{\partial T}{\partial n} \tag{21}$$

Surface-mean Nusselt number is calculated by

$$\overline{Nu} = \frac{1}{A} \int_A NudA \tag{22}$$

where A indicates the surface of the duct. Finally, the time-mean Nusselt number is calculated as

$$\langle \overline{Nu} \rangle = \frac{1}{t} \int_0^t \overline{Nu}dt \tag{23}$$

where t is the time duration.

3.4. Definitions

The following concepts are used in this research:

Reflect boundary condition: This type of boundary usually used as particle boundary condition at wall, symmetry, and axis boundaries [33]. It should be stated that the particle rebounds off of the boundary in regard with a variation in its momentum as specified by the coefficient of restitution. Coefficient of restitution is defined by

$$\text{Coefficient of restitution} = \frac{V_{2,n}}{V_{1,n}} \tag{24}$$

where $V_{1,n}$ and $V_{2,n}$ are the particle velocities before and after particle-wall collision. This boundary has a restitution coefficient of 1.

Trap boundary condition: This type of boundary usually used as particle boundary condition at wall boundaries [33]. The trajectory calculations are terminated and the destiny of the particle is saved as trapped. This boundary has a restitution coefficient of 0.

Escape boundary condition: This type of boundary usually used as a particle boundary condition at all flow boundaries containing pressure and velocity inlets, pressure outlets, etc. [33]. The particle is considered as having "escaped" when it collisions the boundary in question. Trajectory calculations terminate for this type of boundary.

Deposition: The deposition is defined as the ratio of the number of deposited (trapped) particles on all involved surfaces containing obstacle and duct surfaces to the number of particles injected to the duct at specific time.

Concentration (Solid volume fraction): The void fraction of each cell in discrete particle model can be determined by

$$\varepsilon = 1 - \frac{\sum V_i}{\Delta V} \tag{25}$$

where V_i is the volume of ith particle in the cell. Moreover, the summation is taken over all the particles in the cell volume $\Delta V = \Delta x \Delta y d_p$. This means that the two-dimensional domain is regarded as a pseudo three-dimensional one with a thickness of one particle diameter d_p [34].

Parcel: In DPM, it is not possible to track all physical particles. Instead, representative particles or parcels should be tracked. A particle in each parcel is representative of the entire particles on that parcel and motion and heat exchange equations are solved only for this particle and extended to others. After solving the equations for particle, a Gaussian distribution function is used to make the connection between the particle and parcel parameters with

$$\theta_{parcel} = \sum N_{particle} G_w \theta_{particle} \tag{26}$$

$$G_w = \left(\frac{a}{\pi}\right)^{\frac{3}{2}} \exp\left(-a\frac{\left|x_{prcel} - x_{particle}\right|^2}{\Delta x^2}\right) \tag{27}$$

where θ is particle or parcel variable. G_w and N denote the Gaussian function and number of particles.

4. Numerical Results and Discussion

In the current computation, a finite volume method is utilized to discretize the equations of liquid and solid phases. Moreover, the pressure and velocity terms are stored at node center and node faces, respectively by using the staggered grid arrangement. The coupling between the pressure and velocity terms is achieved by SIMPLE algorithm of Patankar [35]. For using SIMPLE algorithm, the mass conservation and momentum equations for fluid are combined to drive the pressure-correction equation. The convection, diffusion, and time terms are discretized by applying a first order upwind scheme, a central difference algorithm, and a first order implicit method, respectively. Finally, the convergence is passed when the values of residuals reduced to $\leq 10^{-4}$ for all equations except the energy equation. This considered value is set at 10^{-6} for energy equation. All simulations are performed by the commercial software Ansys-Fluent. Figure 2 shows the procedure of the numerical solution in Fluent for this problem.

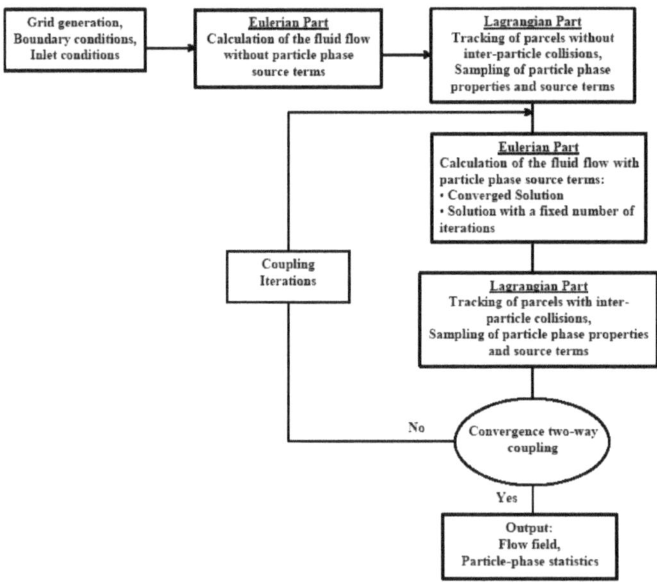

Figure 2. Procedure of numerical solution in Fluent for this problem.

4.1. Grid Study and Validation

A non-uniform square mesh is generated throughout the domain with a more density around the obstacles surfaces. A schematic view of this mesh with a near zone around the obstacles is disclosed in Figure 3. Various mesh sizes are explored to certify the sensitivity of the numerical outputs to the mesh resolution. The results of this test for the mean Nusselt number on the top wall of the duct at $Re = 100$, $S/D = 1.5$, $d_p = 30$ nm, and $\Phi = 0.01$ are presented in Table 1. Note that δ/D in this table indicates the ratio of smallest cell size (δ) to side of the obstacle (D). It is observed that the difference in the mean Nusselt number between Cases 3 and 4 is 0.23%. Hence, the cell number of Case 3 is selected for the subsequent calculations.

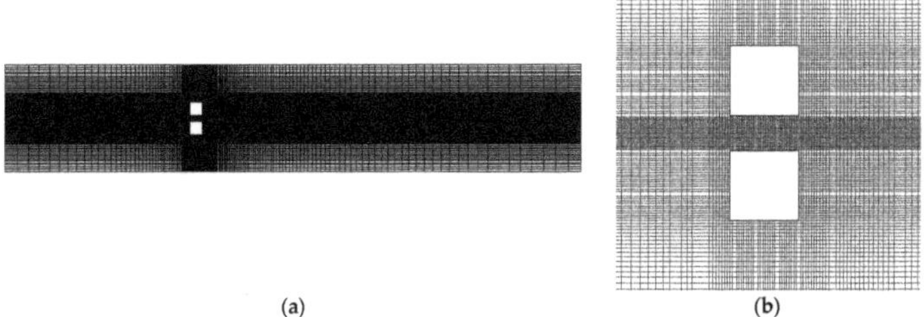

Figure 3. Mesh distribution (**a**) whole domain (**b**) near the obstacles.

Table 1. The grid study at $Re = 100$, $S/D = 1.5$, $d_p = 30$ nm, and $\Phi = 0.01$.

No.	Grid Number	δ/D	$\langle \overline{Nu} \rangle$	Percentage Difference
1	2375	0.05	3.342	1.34%
2	4750	0.033	3.387	1.01%
3	9500	0.025	3.421	0.23%
4	19,000	0.02	3.429	—

To examine the accuracy of the numerical results, the numerical outputs are benchmarked with experimental results obtained by Heyhat et al. [36]. The case for validation is aluminum oxide-water nanofluid flow with $dp = 40$ nm and $\Phi = 0.01$ in a straight tube. The results of comparison between the experimental and numerical results are presented in Figure 4 for the variations of pressure drop ratio with Reynolds number. The pressure drop ratio is defined as the ratio of pressure drop of nanofluid to that of pure water. This comparison showed good agreement between the experimental and numerical results with a relative error about 5%.

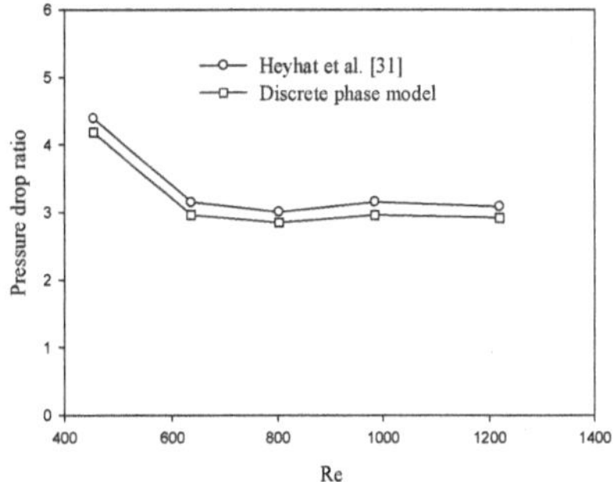

Figure 4. The ratio of pressure drop for Al_2O_3-water nanofluid versus Reynolds numbers at $dp = 40$ nm and $\Phi = 0.01$.

4.2. Discussion

In this part, the results are discussed for various values of parameters containing the space ratio and particle diameter.

The particle dispersions inside the duct for different particle diameters at *Re* = 100, $\Phi = 0.01$, and *S/D* = 3.5 are shown in Figure 5. This figure shows the locations of injected particles at a specific time, which are superimposed on vorticity contours. It is observed that the particles with diameters of 30 nm are dispersed at entire of the duct. Brownian force generated between the liquid molecules and the suspended particles causes the randomized dispersion of the particles. This force is larger in comparison to the inertial forces for smaller values of particle diameter. The particles with smaller diameter (*dp* = 30 nm) are not affected by the flow streamline and they diffuse through the streamlines. However, the particles with larger diameters (i.e., 0.1 and 0.25 μm) accumulate in the vorticity regions around the perimeter of the vortices. It should be stated that the larger particles are under the influence of forces created by the vortices and vortical flow field. Finally, the inertia forces of particles dominate the centrifugal forces formed by the flow for the particles with diameter of 0.5 μm and this causes the exit of particles from the vortices path line and tends them toward the duct surfaces.

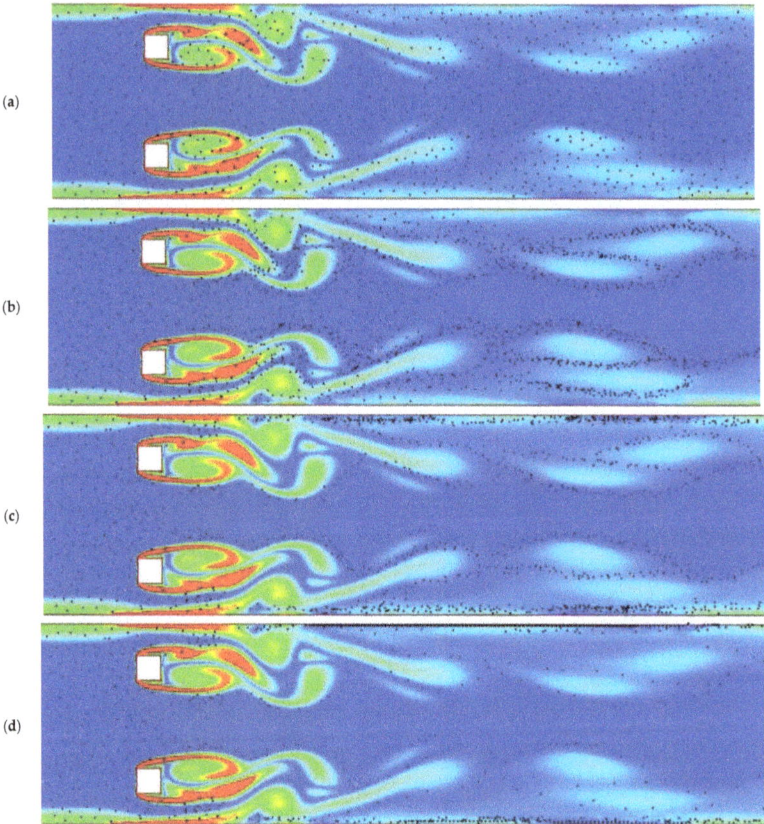

Figure 5. Dispersion of the particles inside the duct for (**a**) *dp* = 30 nm; (**b**) *dp* = 0.1 μm; (**c**) *dp* = 0.25 μm; (**d**) *dp* = 0.5 μm at *Re* = 100, *S/D* = 3.5, and $\Phi = 0.01$.

Figure 6 discloses the concentration contours for different values of space ratio at *Re* = 100, $d_p = 30$ nm, and $\Phi = 0.01$. It is observed that the concentrations of particles for all values of space ratio are almost constant with a very low deposition near the duct surface for the regions before the obstacles. For these regions, the concentration for most parts of the duct keeps constant at the entrance concentration ($\Phi = 0.01$). Note that the particles impacting to the duct walls bounce from the wall due to the reflect assumption of boundary but the flow. Moreover, thermophoretic force for the current case where the liquid temperature is higher than the wall temperature pushes the particles back toward the wall [37]. This causes a deposition of particles near the duct surface. This figure also discloses that the mass diffusion boundary layer is very narrow but is growing along the duct length. Finally, the particles concentration enhances in the recirculating wake region at the downstream of the obstacles. This means that the particles with this size are affected by the flow in the near wake region and some particles are distributed inside the vortex pathways.

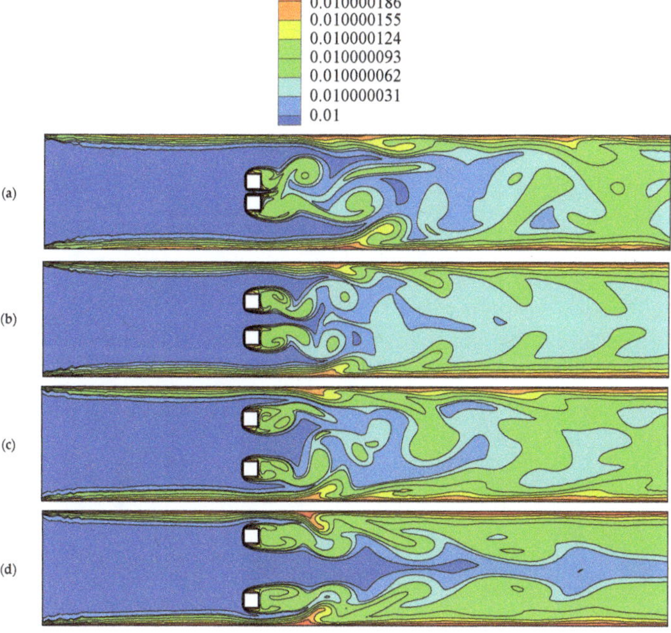

Figure 6. Concentration contours for (**a**) $S/D = 1.5$; (**b**) $S/D = 2.5$; (**c**) $S/D = 3.5$; (**d**) $S/D = 4.5$ at $Re = 100$, d_p =30 nm, and $\Phi = 0.01$.

Effects of particle diameters on the particle deposition at $Re = 100$, $\Phi = 0.01$, and $S/D = 1.5$ are disclosed in Figure 7. The deposition is defined as the ratio of the number of deposited particles on the all involved surfaces containing surfaces of obstacle and duct to the number of particles injected to the duct. It can be observed that the particle deposition enhances with an enhance in the particle diameter. This may be justified by the effect of gravity force on the particle deposition that becomes more significant for higher values of particle diameters. The particle deposition percentages are about 1.1%, 1.8%, 2.7%, and 4.6% for d_p = 30 nm, 0.1 µm, 0.25 µm, and 0.5 µm, respectively. Note that the trap boundary condition with a restitution coefficient of 0 is considered as the particle boundary condition on the surfaces of the duct and obstacles for evaluating particle deposition because for the reflect boundary, the particle deposition is zero as the restitution coefficient is equal to 1.

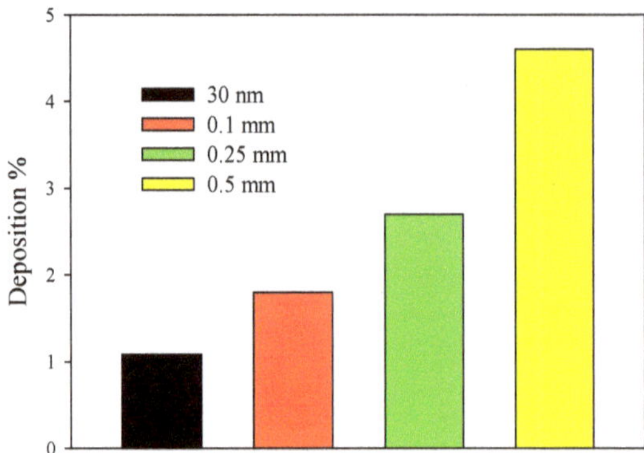

Figure 7. Particle deposition for different particle diameters at Re = 100, Φ = 0.01, and S/D = 1.5.

The effects of space ratio values on the particle deposition at Re = 100, Φ = 0.01, and d_p = 0.1 μm are disclosed in Figure 8. It is observed that the particle deposition enhances with an enhance in the value of space ratio. As mentioned earlier, the particles are affected by the flow in the near wake region for d_p = 0.1 μm. As the space ratio value increases, the distances between obstacles and duct walls decrease and the wake region of each obstacle transfers to the regions near the duct walls. Hence, the number of particles near the center of the duct reduces considerably and this leads to an increase in the number of particles that deposit on the duct's surfaces. The particle deposition percentages are about 1.8%, 2.3%, 3.1%, and 4.3% for S/D = 1.5, 2.5, 3.5, and 4.5, respectively.

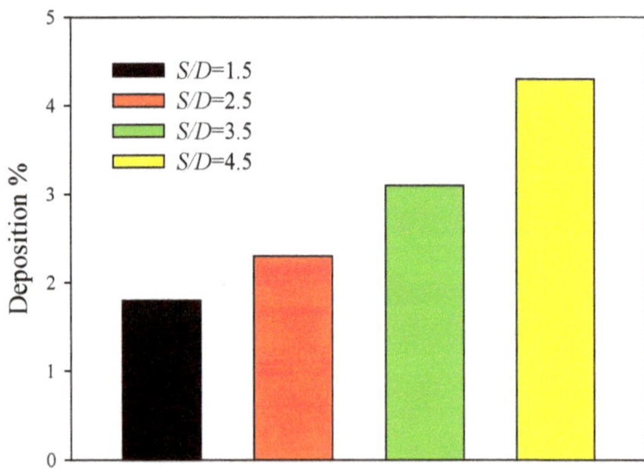

Figure 8. Particle deposition for different values of space ratio at Re = 100, Φ = 0.01, and d_p = 0.1 μm.

Figure 9 discloses the variations of mean Nusselt number on the top wall of the duct with different values of space ratio at Re = 100 and two values of volume fraction of nanoparticles. It can be observed that the mean Nusselt number enhances as the space ratio enhances. Generally, the heat transfer rate enhances by mounting the obstacle inside the duct because mixing of the hot liquid in center of the duct with cold liquid around the wall amends due to the oscillations generated by vortex shedding. As mentioned earlier, the distances between the walls of obstacles and duct decrease with an increase in the space ratio and accordingly, the oscillations generated by vortex shedding transfer to the regions around the duct walls. It worth noting that the fluid mixing near the duct wall increases with an increase in the space ratio, causing a higher temperature gradient around the duct wall and subsequently, a more efficient convection heat transfer. There are about 45% and 41% increments in the mean Nusselt number for $\Phi = 0$ and 0.01, respectively when the space ratio is increased in the range of 1–2.5. Finally, the mean Nusselt number enhances about 10% by using the particles in the liquid with $\Phi = 0.01$ and $d_p = 30$ nm. This can be justified by positive effects of Brownian diffusion of particles and the drag of them on heat transfer improvement. Moreover, the thermal conductivity of nanofluid is more than that of the pure water case and this leads to a higher heat transfer rate.

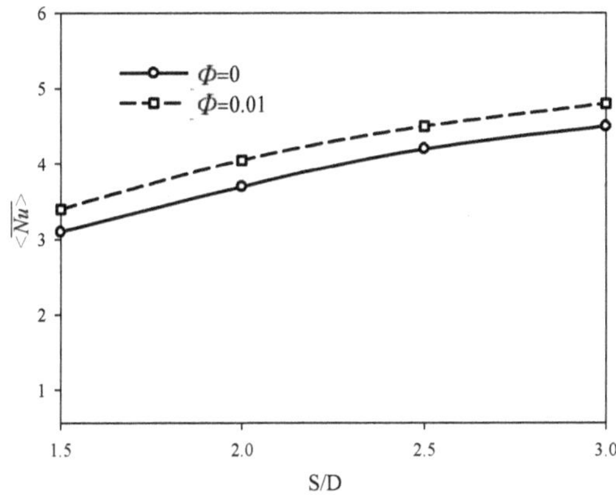

Figure 9. Variation of mean Nusselt number with space ratios for $Re = 100$, $d_p = 30$ nm, and two values of solid volume fraction of nanoparticles.

5. Conclusions

This paper used a two-way coupling of discrete phase model to track the discrete nature of aluminum oxide particles in an obstructed duct with two side-by-side obstacles. The effects of particle diameters and space ratios of obstacles on the dispersion and concentration of particles were evaluated. The obtained results showed that the particles with smaller diameter ($d_p = 30$ nm) are not affected by the flow streamline and they diffuse through the streamlines. However, the particles with larger diameters (i.e., 0.1 μm and 0.25 μm) accumulate in the vorticity regions around the perimeter of the vortices. It was concluded that the particle deposition increases with an increase in the particle diameter. The particle deposition percentages are about 1.1%, 1.8%, 2.7%, and 4.6% for $d_p = 30$ nm, 0.1 μm, 0.25 μm, and 0.5 μm, respectively. Moreover, about 45% and 41% increments in the mean Nusselt number were observed for $\Phi = 0$ and 0.01, respectively when the space ratio is increased in the range of 1–2.5. Finally, it was found that the mean Nusselt number enhances about 10% by using the particles in the liquid with $\Phi = 0.01$ and $d_p = 30$ nm.

Acknowledgments: R. Ellahi is grateful to PCST, Ministry of Science and Technology Pakistan to honed him with 5th top most Productive Scientist Award in category A in all subject for the year 2016.

Conflicts of Interest: The authors declare no conflict of interest.

References

1. Mahir, N.; Altaç, Z. Numerical investigation of convective heat transfer in unsteady flow past two cylinders in tandem arrangements. *Int. J. Heat Fluid Flow* **2008**, *29*, 1309–1318. [CrossRef]
2. Rashidi, M.M.; Abelman, S.; Freidoonimehr, N. Entropy generation in steady MHD flow due to a rotating porous disk in a nanofluid. *Int. J. Heat Mass Transf.* **2013**, *62*, 515–525. [CrossRef]
3. Sekrani, G.; Poncet, S. Further investigation on laminar forced convection of nanofluid flows in a uniformly heated pipe using direct numerical simulations. *Appl. Sci.* **2016**, *6*, 1–24. [CrossRef]
4. Vanaki, S.M.; Ganesan, P.; Mohammed, H.A. Numerical study of convective heat transfer of nanofluids: A review. *Renew. Sustain. Energy Rev.* **2016**, *54*, 1212–1239. [CrossRef]

5. He, Y.; Men, Y.; Zhao, Y.; Lu, H.; Ding, Y. Numerical investigation into the convective heat transfer of TiO_2 nanofluids flowing through a straight tube under the laminar flow conditions. *Appl. Therm. Eng.* **2009**, *29*, 1965–1972. [CrossRef]
6. Mirzaei, M.; Saffar-Avval, M.; Naderan, H. Heat transfer investigation of laminar developing flow of nanofluids in a microchannel based on Eulerian–Lagrangian approach. *Appl. Therm. Eng.* **2014**, *92*, 1139–1149. [CrossRef]
7. Bahremand, H.; Abbassi, A.; Saffar-Avval, M. Experimental and numerical investigation of turbulent nanofluid flow in helically coiled tubes under constant wall heat flux using Eulerian–Lagrangian approach. *Powder Technol.* **2015**, *269*, 93–100. [CrossRef]
8. Turki, S.; Abbassi, H.; Nasrallah, S.B. Two-dimensional laminar fluid flow and heat transfer in a channel with a built-in heated square cylinder. *Int. J. Therm. Sci.* **2003**, *42*, 1105–1113. [CrossRef]
9. Pirouz, M.M.; Farhadi, M.; Sedighi, K.; Nemati, H.; Fattahi, E. Lattice Boltzmann simulation of conjugate heat transfer in a rectangular channel with wall-mounted obstacles. *Sci. Iran. B* **2011**, *18*, 213–221. [CrossRef]
10. Heidary, H.; Kermani, M.J. Heat transfer enhancement in a channel with block(s) effect and utilizing Nano-fluid. *Int. J. Therm. Sci.* **2012**, *57*, 163–171. [CrossRef]
11. Mohyud-Din, S.T.; Khan, U.; Ahmed, N.; Hassan, S.M. Magnetohydrodynamic flow and heat transfer of nanofluids in stretchable convergent/divergent channels. *Appl. Sci.* **2015**, *5*, 1639–1664. [CrossRef]
12. Selimefendigil, F.; Öztop, H. Pulsating nanofluids jet impingement cooling of a heated horizontal surface. *Int. J. Heat Mass Transf.* **2014**, *69*, 54–65. [CrossRef]
13. Khan, W.; Gul, T.; Idrees, M.; Islam, S.; Khan, I.; Dennis, L.C.C. Thin filmWilliamson nanofluid flow with varying viscosity and thermal conductivity on a time-dependent stretching sheet. *Appl. Sci.* **2016**, *6*, 334. [CrossRef]
14. Sheikholeslami, M.; Ellahi, R. Electrohydrodynamic nanofluid hydrothermal treatment in an enclosure with sinusoidal upper wall. *Appl. Sci.* **2015**, *5*, 294–306. [CrossRef]
15. Akbar, N.S. Endoscopy analysis for the peristaltic flow of nanofluids containing carbon nanotubes with heat transfer. *Z. Naturforschung A* **2015**, *70*, 745–755. [CrossRef]
16. Selimefendigil, F.; Öztop, H. MHD mixed convection of nanofluid filled partially heated triangular enclosure with a rotating adiabatic cylinder. *J. Taiwan Inst. Chem. Eng.* **2014**, *45*, 2150–2162. [CrossRef]
17. Sheikholeslami, M.; Zia, Q.M.; Ellahi, R. Effect of induced magnetic field on free convective heat transfer of nanofluid considering KKL correlation. *Appl. Sci.* **2016**, *6*, 324. [CrossRef]
18. Bhatti, M.M.; Abbas, T.; Rashidi, M.M.; Ali, M.E.; Yang, Z. Entropy generation on MHD Eyring-Powell nanofluid through a permeable stretching surface. *Entropy* **2016**, *18*, 224. [CrossRef]
19. Mahian, O.; Kianifar, A.; Kalogirou, S.A.; Pop, I.; Wongwises, S. A review of the applications of nanofluids in solar energy. *Int. J. Heat Mass Transf.* **2013**, *57*, 582–594. [CrossRef]
20. Ahmad, S.; Rohni, A.M.; Pop, I. Blasius and sakiadis problems in nanofluids. *Acta Mech.* **2011**, *218*, 195–204. [CrossRef]
21. Shahmohamadi, H.; Rahmani, R.; Rahnejat, H.; Garner, C.P.; Balodimos, N. Thermohydrodynamics of lubricant flow with carbon nanoparticles in tribological contacts. *Tribol. Int.* **2016**. [CrossRef]
22. Shahmohamadi, H.; Rashidi, M.M. VIM solution of squeezing MHD nanofluid flow in a rotating channel with lower stretching porous surface. *Adv. Powder Technol.* **2016**, *27*, 171–178. [CrossRef]
23. Cheng, W.L.; Sadr, R. Induced flow field of randomly moving nanoparticles: A statistical perspective. *Microfluid Nanofluid* **2015**, *18*, 1317–1328. [CrossRef]
24. Bovand, M.; Rashidi, S.; Ahmadi, G.; Esfahani, J.A. Effects of trap and reflect particle boundary conditions on particle transport and convective heat transfer for duct flow—A two-way coupling of Eulerian–Lagrangian model. *Appl. Therm. Eng.* **2016**, *108*, 368–377. [CrossRef]
25. Bianco, V.; Chiacchio, F.; Manca, O.; Nardini, S. Numerical investigation of nanofluids forced convection in circular tubes. *Appl. Therm. Eng.* **2009**, *29*, 3632–3642. [CrossRef]
26. Talbot, L. Thermophoresis of particles in a heated boundary layer. *J. Fluid Mech.* **1980**, *101*, 737–758. [CrossRef]
27. Fluent Inc. *Fluent 6.2 User Manual*; Fluent Incorporated: Lebanon, NH, USA, 2006.
28. Minkowycz, W.J.; Sparrow, E.M.; Murthy, J.Y. *Handbook of Numerical Heat Transfer*, 2nd ed.; John Wiley & Sons: Hoboken, NJ, USA, 2006.

29. Garoosi, F.; Shakibaeinia, A.; Bagheri, G. Eulerian–Lagrangian modeling of solid particle behavior in a square cavity with several pairs of heaters and coolers inside. *Powder Technol.* **2015**, *280*, 239–255. [CrossRef]
30. Li, A.; Ahmadi, G. Dispersion and deposition of spherical particles from point sources in a turbulent channel flow. *Aerosol Sci. Technol.* **1992**, *16*, 209–226. [CrossRef]
31. Ounis, H.; Ahmadi, G.; McLaughlin, J.B. Brownian diffusion of submicrometer particles in the viscous sublayer. *J. Colloid Interface Sci.* **1991**, *143*, 266–277. [CrossRef]
32. Ranz, W.E., Jr.; Marshall, W.R. Evaporation from drops, Part I. *Chem. Eng. Prog.* **1952**, *48*, 141–146.
33. Fluent Inc. *FLUENT15 User Manual*; Fluent Incorporated: Lebanon, NH, USA, 2014.
34. Wu, C.L.; Berrouk, A.S. Discrete particle model for dense gas-solid flows. *Adv. Multiph. Flow Heat Transf.* **2012**, *3*, 151–187.
35. Patankar, S.V. *Numerical Heat Transfer and Fluid Flow*; Hemisphere: New York, NY, USA, 1980.
36. Heyhat, M.M.; Kowsary, F.; Rashidi, A.M.; Momenpour, M.H.; Amrollahi, A. Experimental investigation of laminar convective heat transfer and pressure drop of water-based Al_2O_3 nanofluids in fully developed flow regime. *Exp. Therm. Fluid Sci.* **2013**, *44*, 483–489. [CrossRef]
37. Rashidi, S.; Bovand, M.; Esfahani, J.A.; Ahmadi, G. Discrete particle model for convective Al_2O_3-water nanofluid around a triangular obstacle. *Appl. Therm. Eng.* **2016**, *100*, 39–54. [CrossRef]

Editorial

Special Issue on Recent Developments of Nanofluids

Rahmat Ellahi [1,2]

[1] Department of Mathematics & Statistics, International Islamic University (IIUI), Islamabad 44000, Pakistan; rahmatellahi@yahoo.com or rellahi@alumni.ucr.edu

[2] Department of Mechanical Engineering, University of California Riverside, Riverside, CA 92521, USA

Received: 16 January 2018; Accepted: 23 January 2018; Published: 27 January 2018

1. Introduction

Recent advances in nanotechnology have allowed the development of a new category of fluids termed nanofluids. A nanofluid refers to the suspension of nanosize particles, which are suspended in the base fluid with low thermal conductivity. The base fluid, or dispersing medium, can be aqueous or non-aqueous in nature. Typical nanoparticles are metals, oxides, carbides, nitrides, or carbon nanotubes. These shapes may be spheres, disks, rods, etc. By using these additives, one can increase the heat transfer coefficient and consequently enhance the heat transfer value and performance of base fluids. Some of these fluids can be considered Newtonian fluids, but in many applications the Newtonian model is not very accurate; therefore, it has generally been acknowledged that non-Newtonian fluids exhibiting a nonlinear relationship between the stresses and the rate of strain are more appropriate in technological applications as compared to Newtonian fluids. Many industrial fluids are non-Newtonian in their flow characteristics and are referred to as rheological fluids, such as slurries (china clay and coal in water, sewage sludge, etc.) and multiphase mixtures (oil-water emulsions, gas-liquid dispersions, such as froths and foams, butter). Further examples displaying a variety of non-Newtonian characteristics include pharmaceutical formulations, cosmetics and toiletries, paints, synthetics lubricants, biological fluids (blood, synovial fluid, salvia), and food stuffs (jams, jellies, soups, marmalades), etc. Moreover, simulation of boundary layer flow of nanofluids is another aspect of this special issue that has various applications in engineering and industrial disciplines.

Existing literature indicates that despite a vast range of application, the investigation on proposed title was still scant. Consequently, researchers are invited to contribute their original research and review articles with this hope that this special issue will also serve as a forum for presenting innovative and new developments quite relevant for the scope of this special issue as specified in keywords.

A total of 12 papers were submitted for possible publication in this special issue. After a comprehensive peer review, only eight papers qualified to get the acceptance for final publication. The rest of papers could not be accommodated. The submissions were technically correct, but were not considered appropriate for the scope of this special issue. The authors are from geographically distributed countries such as China, Romania, South Africa, Iran, Pakistan, Malaysia, and Saudi Arabia. This reflects the great impact of the proposed topic and the effective organization of the guest editorial team of this special issue.

2. Nanofluids: Techniques and Applications

The effect of thermal radiation on the thin film nanofluid flow of a Williamson fluid over an unsteady stretching surface with variable fluid properties is investigated in [1]. Special attention has been given to the variable fluid properties. Analytical solutions of nonlinear governing equations are achieved by means of homotopy analysis method. Experimental values of the Prandtl number have been used to produce the results for the Williamson nanofluid, whereas the accuracy of the HAM results has been verified via numerical solutions. The effects of non-dimensional physical parameters—such as thermal conductivity, Schmidt number, Williamson parameter, Brinkman number, radiation parameter, and Prandtl number—have been thoroughly demonstrated and discussed. A comparison is also made for the validation of obtained results.

In the paper "On Squeezed Flow of Jeffrey Nanofluid between Two Parallel Disks", Hayat et al. [2] presented the magnetohydrodynamic (MHD) squeezing flow of Jeffrey nanofluid between two parallel disks. Constitutive relations of Jeffrey fluid are employed in the problem development. Heat and mass transfer aspects are examined in the presence of thermophoresis and Brownian motion. Jeffrey fluid subject to time dependent applied magnetic field is conducted. Suitable variables lead to a strong nonlinear system. The resulting systems are computed via a homotopic approach. The behaviors of several pertinent parameters are analyzed through graphs and numerical data. Skin friction coefficient and heat and mass transfer rates are numerically examined. They found that the larger values of Deborah numbers correspond to lower temperature and concentration profiles. Both temperature and concentration profiles are higher for larger values of thermophoresis parameter. The present analysis reduces to a Newtonian nanofluid flow situation as a limiting case of this model.

It is well known that the best way of convective heat transfer is the flow of nanofluids through a porous medium. In this regard, a mathematical model is presented in [3] to study the effects of variable viscosity, thermal conductivity and slip conditions on the steady flow and heat transfer of nanofluids over a porous plate embedded in a Darcy-type porous medium. The nanofluid viscosity and thermal conductivity are assumed to be linear functions of temperature, and the wall slip conditions are employed in terms of shear stress. The similarity transformation technique is used to reduce the governing system of partial differential equations to a system of nonlinear ordinary differential equations. The resulting system of ODEs is then solved numerically using the shooting technique. The numerical values obtained for the velocity and temperature profiles, skin friction coefficient, and Nusselt number are presented and discussed through graphs and tables. It is shown that the increase in the permeability of the porous medium, the viscosity of the nanofluid, and the velocity slip parameter decrease the momentum and thermal boundary layer thickness and eventually increase the rate of heat transfer. Moreover, the analysis can be extended to include the results for different water-based nanofluids, and a comparison can be generated on the heat transfer characteristics of different nanofluids. Clearly, there is an opportunity to consider/extend this problem with non-Newtonian nanofluid models and to perform experimental work on these systems.

The studies of classical nanofluids are restricted to models described by partial differential equations of integer order, and the memory effects are ignored. Fractional nanofluids, modeled by differential equations with Caputo time derivatives, are able to describe the influence of memory on the nanofluid behavior. In the paper [4], the heat and mass transfer characteristics of two water-based fractional nanofluids, containing nanoparticles of CuO and Ag, over an infinite vertical plate with a uniform temperature and thermal radiation, are analytically and graphically studied. Closed form solutions are determined for the dimensionless temperature and velocity fields, as well as the corresponding Nusselt number and skin friction coefficient. These solutions, presented in equivalent forms in terms of the Wright function or its fractional derivatives, have also been reduced to the known solutions of ordinary nanofluids. The influence of the fractional parameter on the temperature, velocity, Nusselt number, and skin friction coefficient is graphically underlined and discussed. The enhancement of heat transfer in the natural convection flows is lower for fractional nanofluids, in comparison to ordinary nanofluids. In both cases, the fluid temperature increases for increasing values of the nanoparticle volume fraction.

The magnetohydrodynamic thin film nanofluid sprayed on a stretching cylinder with heat transfer is explored in [5]. The spray rate is a function of film size. Constant reference temperature is used for the motion past an expanding cylinder. The sundry behavior of the magnetic nano-liquid thin film is carefully noticed which results in to bring changes in the flow pattern and heat transfer. Water-based nanofluids like Al_2O_3-H_2O and CuO-H_2O are investigated under the consideration of thin film. The basic constitutive equations for the motion and transfer of heat of the nanofluid with the boundary conditions have been converted to nonlinear coupled differential equations with physical conditions by employing appropriate similarity transformations. The modeled equations have been solved using homotopic approach and lead to detailed expressions for the velocity profile and temperature distribution. The pressure distribution and spray rate are also calculated. The residual errors show the authentication of the present work. The CuO-H_2O nanofluid results from this study are compared with the experimental results reported in the literature and are found to have excellent agreement. The present work discusses the salient features of all the indispensable parameters of spray rate, velocity profile, temperature, and pressure distributions which have been displayed graphically and illustrated.

In [6], the Brownian motion and thermophoresis effect of the liquid film flow over an unstable stretching surface in a porous space is presented. The main focus is on the variation in the thickness of the liquid film in a porous space. The graphical comparison of these two methods is elaborated. The physical and numerical results using h curves for the residuals of the velocity, temperature, and concentration profiles are obtained. The key observations concluded that higher values of porosity parameter generates larger open space and create hurdle to flow and as a result the flow field reduces. The larger values of Prandtl number reduces the thermal boundary layer due to which the temperature field reduces while the Eckert number is allied with the viscous dissipation term and lead to an increase the quantity of heat being produced by the shear forces in the fluid. Therefore, larger values of Eckert raise the temperature field. It is also seen that the larger values of thickness parameter transport more fluid in the boundary layer region and cooling effect is produced which absorbs the heat transfer from the sheet and, as a result, the temperature reduces.

A radial basis function (RBF) neural network with three-layer feed forward architecture was developed in [7] to effectively predict the viscosity ratio of different ethylene glycol/water based nanofluids. A total of 216 experimental data involving CuO, TiO_2, SiO_2, and SiC nanoparticles were collected from the published literature to train and test the RBF neural network. The parameters—including temperature, nanoparticle properties (size, volume fraction, and density), and viscosity of the base fluid—were selected as the input variables of the RBF neural network. The investigations demonstrated that the viscosity ratio predicted by the RBF neural network agreed well with the experimental data. In addition, by comparing the RBF predictive values and the experimental data published in various studies, it was demonstrated that the RBF neural network not only exhibited good modeling accuracy but also could effectively predict the influences of temperature, nanoparticle volume fraction, and diameter on the viscosity ratio of different ethylene glycol/water based nanofluids. Compared to the typical viscosity models, namely the Batchelor model and Chen model, the RBF neural network has a good ability to predict the viscosity ratio of different ethylene glycol/water based nanofluids. However, the prediction performance can be affected by the size of the data set. The present investigation may play an active role for developing the modeling of nanofluid viscosity. However, how to extend the application of ANN to predict other thermo-physical properties of nanofluids is still worthy of study in the future.

A two-way coupling of discrete phase model is developed in order to track the discrete nature of aluminum oxide particles in an obstructed duct with two side-by-side obstacles is explored in [8]. Finite volume method and trajectory analysis are simultaneously utilized to solve the equations for liquid and solid phases, respectively. The interactions between two phases are fully taken into account in the simulation by considering the Brownian, drag, gravity, and thermophoresis forces. The effects of space ratios between two obstacles and particle diameters on different parameters containing concentration and deposition of particles and Nusselt number are studied for the constant values of the Reynolds numbers and volume fractions of nanoparticles. The obtained results indicate that the particles with smaller diameter (dp = 30 nm) are not affected by the flow streamline and they diffuse through the streamlines. Moreover, the particle deposition is enhanced as the value of the space ratio increases. A comparison between the experimental and numerical results is also provided with the existing literature as a limiting case of the reported problem and found in good agreement.

3. Future Trends in Nanotechnology

Even as with the completion of this special issue, the material that advances the state-of-the-art of experimental, numerical, and theoretical methodologies or extends the bounds of existing methodologies to new contributions in applied nano-technology is still insufficient. Nanofluids strengthen solar energy applications such as heat exchanger design and medical applications including cancer therapy and safer surgery by heat treatment. Nanofluid technology can also help to develop better oils and lubricants for practical applications.

Acknowledgments: The guest editorial team of Applied Sciences would like to thank all authors for contributing their original work to this special issue, no matter what the final decisions of the submitted manuscripts were. The editorial team would also like to thank all anonymous professional reviewers for their valuable time, comments, and suggestions during the review process. We also acknowledge the entire staff of the journal's editorial board for providing their cooperation regarding this special issue. We hope that the scientists who are working in the same regime will not only enjoy this special issue, but would also appreciate the efforts made by the entire team.

Conflicts of Interest: The author declares no conflict of interest.

References

1. Khan, W.; Gul, T.; Idrees, M.; Islam, S.; Khan, I.; Dennis, L.C.C. Thin Film Williamson Nanofluid Flow with Varying Viscosity and Thermal Conductivity on a Time-Dependent Stretching Sheet. *Appl. Sci.* **2016**, *6*, 334. [CrossRef]
2. Hayat, T.; Abbas, T.; Ayub, M.; Muhammad, T.; Alsaedi, A. On Squeezed Flow of Jeffrey Nanofluid between Two Parallel Disks. *Appl. Sci.* **2016**, *6*, 346. [CrossRef]
3. Hussain, S.; Aziz, A.; Aziz, T.; Khalique, C.M. Slip Flow and Heat Transfer of Nanofluids over a Porous Plate Embedded in a Porous Medium with Temperature Dependent Viscosity and Thermal Conductivity. *Appl. Sci.* **2016**, *6*, 376. [CrossRef]
4. Fetecau, C.; Vieru, D.; Azhar, W.A. Natural Convection Flow of Fractional Nanofluids over an Isothermal Vertical Plate with Thermal Radiation. *Appl. Sci.* **2017**, *7*, 247. [CrossRef]
5. Khan, N.S.; Gul, T.; Islam, S.; Khan, I.; Alqahtani, A.M.; Alshomrani, A.S. Magnetohydrodynamic Nanoliquid Thin Film Sprayed on a Stretching Cylinder with Heat Transfer. *Appl. Sci.* **2017**, *7*, 271. [CrossRef]
6. Ali, L.; Islam, S.; Gul, T.; Khan, I.; Dennis, L.C.C.; Khan, W.; Khan, A. The Brownian and Thermophoretic Analysis of the Non-Newtonian Williamson Fluid Flow of Thin Film in a Porous Space over an Unstable Stretching Surface. *Appl. Sci.* **2017**, *7*, 404. [CrossRef]
7. Zhao, N.; Li, Z. Viscosity Prediction of Different Ethylene Glycol/Water Based Nanofluids Using a RBF Neural Network. *Appl. Sci.* **2017**, *7*, 409. [CrossRef]
8. Rashidi, S.; Esfahani, J.A.; Ellahi, R. Convective Heat Transfer and Particle Motion in an Obstructed Duct with Two Side by Side Obstacles by Means of DPM Model. *Appl. Sci.* **2017**, *7*, 431. [CrossRef]

MDPI
St. Alban-Anlage 66
4052 Basel
Switzerland
Tel. +41 61 683 77 34
Fax +41 61 302 89 18
www.mdpi.com

Applied Sciences Editorial Office
E-mail: applsci@mdpi.com
www.mdpi.com/journal/applsci

www.ingramcontent.com/pod-product-compliance
Lightning Source LLC
Chambersburg PA
CBHW041218220326
41597CB00033BA/6031

* 9 7 8 3 0 3 8 4 2 8 3 3 6 *